Origins of Mathematical Words

Origins of Mathematical Words

A Comprehensive Dictionary of Latin, Greek, and Arabic Roots

Anthony Lo Bello

The Johns Hopkins University Press
Baltimore

© 2013 The Johns Hopkins University Press
All rights reserved. Published 2013
Printed in the United States of America on acid-free paper
2 4 6 8 9 7 5 3 1

The Johns Hopkins University Press
2715 North Charles Street
Baltimore, Maryland 21218-4363
www.press.jhu.edu

Library of Congress Cataloging-in-Publication Data

Lo Bello, Anthony, 1947–
Origins of mathematical words : a comprehensive dictionary of Latin, Greek,
and Arabic roots / by Anthony Lo Bello.
pages cm
Includes bibliographical references.
ISBN-13: 978-1-4214-1098-2 (pbk. : alk. paper)
ISBN-10: 1-4214-1098-2 (pbk. : alk. paper)
ISBN-13: 978-1-4214-1099-9 (electronic)
ISBN-10: 1-4214-1099-0 (electronic)
1. Mathematics–Terminology.   I. Title.
QA41.3.B45 2013
510.1'4–dc23   2013005022

A catalog record for this book is available from the British Library.

Special discounts are available for bulk purchases of this book. For more information, please contact Special Sales at 410-516-6936 or specialsales@press.jhu.edu.

The Johns Hopkins University Press uses environmentally friendly book materials, including recycled text paper that is composed of at least 30 percent post-consumer waste, whenever possible.

Contents

Preface

This is a book about words, mathematical words, how they are made and how they are used. If one admits the proverb that life without literature is death, then one must agree that the correct formation and use of words is essential for any literature, whether mathematical or otherwise.

> If the way in which men express their thoughts is slipshod and mean, it will be very difficult for their thoughts themselves to escape being the same. (Henry Alford, *A Plea for the Queen's English: Stray Notes on Speaking and Spelling*, tenth thousand, Alexander Strahan publisher, London and New York, 1866, pp. 5–6)

In October 2008, Trevor Lipscombe, at the time editor-in-chief of the Johns Hopkins University Press, suggested to me that I undertake to write what he called *a discursive etymological dictionary of mathematical words* whose origins were in Greek, Latin, or Arabic, that is to say, in those languages that I have studied sufficiently so as to be able to comment on the derivations of words that proceeded from them.

There are other dictionaries of mathematical words. That of James and James (*Mathematics Dictionary*, van Nostrand Reinhold, New York, 1959) is justly famous, but it is not an etymological dictionary, so the reader will find little in these pages that might already have been discovered in theirs. The valuable work of Schwartzman (*The Words of Mathematics*, Mathematical Association of America, 1994) may be consulted with benefit by anyone who looks into this book, and such an investigator will notice the ways in which I differ from my learned colleague: I have retained the Greek and Arabic alphabets to avoid the dark and doubtful consequences of transliteration, I have sat in judgment on the correctness of the words I explain, and I have used my license to be discursive to discuss not only the function of mathematics in liberal education but also English usage among mathematicians and their colleagues in the learned world. Since the majority of mathematicians earn their living on the faculties of colleges and universities, I have further commented on the use of words and the style of prose to be found nowadays in these establishments, and which mathematicians for the most part have adopted in their bureaucratic activities such as committee reports, minutes, departmental newsletters, and discussions about mathematics education and curriculum.

Although Herodotus assures us that mathematics, like Egypt, was the gift of the Nile, the Egyptian language had no influence on subsequent mathematical vocabulary. Neither did the inhabitants of Mesopotamia employ any word that survives in modern mathematical usage. The Greeks, as the word *mathematics* itself testifies, were the people responsible for developing our subject as the system of consecutive thought as we know it today, and it is to their language that the earliest mathematical words still in use are to be traced. As *mathematics* is a Greek word, so the earliest mathematical vocabulary was Greek. The mathematical vocabulary of the Greeks has for two thousand years been the common patrimony of our science. It was among the Greeks that the principle *ars gratia artis* was first applied to mathematics; it is a principle on which the chief of the *philosophes* commented disapprovingly in the book in which he introduced Newton to the continent:

...Tous les arts sont à peu près dans ce cas; il y a un point, passé lequel les recherches ne sont plus que pour la curiosité: ces verities ingénieuses et inutiles ressemblent à des étoiles qui, places trop loin de nous, ne nous donnent point de claret. (Voltaire, *Lettres philosophiques, ou Lettres anglaises*, Éditions Garnier Frères, Paris, 1964, vingt-quatrième lettre, p. 139)

...This is very nearly the case with most of the arts: there is a certain point beyond which all researches serve to no other purpose than merely to delight an inquisitive mind. Those ingenious and useless truths may be compared to stars which, by being placed at too great a distance, cannot afford us the least light. (Translation found in the Harvard Classics, Easton Press Millennium Edition, vol. 34, *French and English Philosophers*, p. 162)

During the period of the Roman Empire, some of the Greek mathematical literature was translated into the Latin tongue, although the most common practice was to study the subject in the language in which it was written and even to travel to Greece to do so, as Horace (*Epistiolarum liber* II 2, 41–45) testified:

> Romae nutriri mihi contigit atque doceri,
> Iratus Grais quantum nocuisset Achilles.
> Adiecere bonae paulo plus artis Athenae,
> Scilicet ut vellem curvo dinoscere rectum
> Atque inter silvas Academi quaerere verum.

> It was my Fortune to be bred and taught
> At Rome, what Woes enrag'd Achilles wrought
> To Greece: kind Athens yet improv'd my Parts
> With some small Tincture of ingenuous Arts,
> To learn a right Line from a Curve, and rove
> In search of Wisdom through the museful Grove.
> (Translation by Francis)

The chief early Latin translation of a Greek mathematical text was the edition by Boëthius of the *Elements* of Euclid, accomplished in the late fifth century A.D., shortly after the fall of Rome. Boëthius transliterated rather than translated some of the Greek technical terms, such as *basis, diameter, gnomon, isosceles, orthogonal, parallel, parallelogram, rhomboid, rhombus, scalene,* and *trapezia;* other transliterations of his did not survive the passage of time, for example, *aethimata* (postulates) *amblygonium* (obtuse-angled), *oxygonium* (acute-angled), and *cynas etnyas* (common notions or axioms). Other Greek terms he actually translated into Latin, thereby producing the ancestors of our current English words: *acutus, aequiangulus, aequilaterus, aequus, alternus, angulus, circulus, circumferens, componens, contactus, describere, dividere, exterior, extremus, figura, incidere, infinitus, interior, linea, magnitudo, multilaterus, multiplicare, obtusus, perpendicularis, planus, portio, proportio, punctum, quadrilaterus, recta (right), rectiangulus, rectilineus, secans, sectio, sector, semicirculus, spatium, subtendere, superficies (surface), supplementum, tangens, trilaterus, vertex.*

When the conquests of Islam brought the Arabs into contact with the Byzantine Empire, the caliph requested manuscripts of scientific knowledge from the emperor at Constantinople, and the text of Euclid was introduced to the Muslims. The translations of the Greek texts into the Arabic language were the productions of learned authorities, among whom may be mentioned al-Hajjaj and Ishaq, the translators of Euclid. The following introduction to the commentary of al-Nayrizi (died *circa* 922) on Euclid's *Elements* tells how this was done:

> In the name of Allah, the compassionate, the merciful! Praise be unto Allah, Lord of the worlds, and may Allah be gracious unto Mohammed and unto his family, all of them.
>
> This is the abridgment of the book of Euclid on the study of the *Elements* preliminary to the study of plane geometry, just as the study of the letters of the alphabet, which are the elements of composition, are preliminary to composition. This is the book which Yahya bin Khalid bin Barmak ordered to be translated from the Roman tongue [that is, Greek] into the Arabic tongue at the hands of al-Hajjaj bin Yusuf Matar. And when Allah brought into his caliphate the Imam Mamun Abdullah bin Harun, the Commander of the Faithful, who

delighted in learning and was enthusiastic about wisdom, who was close to scholars and beneficent unto them, al-Hajjaj bin Yusuf saw that he could find favor with him by correcting this book, by summing it up, and by abbreviating it. And so there was left nothing superfluous in it that he did not make succinct, nor any flaw that he did not fix, nor any defect that he did not set aright and rectify, until he had corrected it, made it certain, summed it up, and abbreviated it for people of understanding, discrimination, and learning, without his having changed any of its meaning at all. And he left the earlier edition as it stood for the public. Then Abu'l Abbas bin Hatim al-Nayrizi wrote a commentary upon it, revised some of its formulations, and expanded every part over and above the words of Euclid with what was fitting from the works of others, from the former geometricians, and from the works of them that had commented on the book of Euclid. (Anthony Lo Bello, *The Commentary of al-Nayrizi on Book I of Euclid's Elements of Geometry, with an Introduction on the Transmission of Euclid's Elements in the Middle Ages,* Brill Academic Publishers, Inc., Boston and Leiden, 2003, p. 25)

The words of Arabic origin that entered mathematics at this time include the ancestors of our *algebra, algorithm, azimuth, zenith,* and *zero.*

At the time of the Crusades, the intercourse between Western Europeans and the followers of Islam had the pleasant result that certain qualified scholars like Adelard of Bath translated the Arabic translations of Greek mathematics into Latin, an enterprise that led to the renewal of science in Western Europe.

And if anyone should demand an explanation of all the matters so simply expounded above, he should know that such an explanation must be formulated from Euclid's fifteen books of the geometrical art which we have translated from the Arabic into the Latin language. (My translation of a passage in Adelard's *De Opere Astrolapsus,* to be found on p. 20 in M. Clagett, "The Mediaeval Latin Translations from the Arabic of the *Elements* of Euclid, with Special Emphasis on the Versions of Adelard of Bath," *Isis* 44 [1953], pp. 16–42.)

Here begins the foundation-work of the geometrical art as described by Euclid in fifteen books, translated from the Arabic into the Latin language by Adelard of Bath. (Translation from the Title to MS Oxford, Trinity College 47)

Among the words that we find in Adelard that are not in Boëthius, and whose use has become standard in our day, are: *applicatio, assignatus, coalterni anguli, contactus, demonstrare, equidistans, expansio, extremitates, extrinsecus, intrinsecus, protrahere.* Other translators followed Adelard, and the most widespread medieval Latin edition of Euclid made from the Arabic was a compilation by Robert of Chester that drew from both Boëthius and Adelard.

Eventually, at the time of the Renaissance, the Latin editions of the Greek mathematicians could be made directly from the Greek manuscripts, without the mediation of the Arabic language. Latin remained the main language of mathematical activity in Europe until the nineteenth century, when it at last gave way to the major modern European languages. During this period, the mathematical technical vocabulary was increased by the addition of Latin words that are the parents of words used every day by students of mathematics in American colleges, and of which we may mention the following examples: *abscissa, additio, calculus, calculus integralis, coefficiens, cosecans, cosinus, curva, cyclois, differentiare, divisio, exponentialis, formula, functio, maxima, minima, multiplicatio, ordinata, probabilitas, secans, series, subtractio, tangens.* These words were the invention of authors like the Bernoullis, Euler, Leibniz, and Newton, who were masters of Latin and knew what they were doing when they coined new words.

The words that have come into use since the disappearance of Latin from the curriculum of general education, that is to say, those that became current in the twentieth and twenty-first centuries, exhibit as a rule more of the peculiarities of concoctions that Dr. Johnson called *low.* Such compositions are frequently acronyms or *macaronic* concatenations, the infallible sign of defective education. Words that fall into this category are *analog, ANOVA, antichain, approximoscope, autocorrelation, biholomorphic, cohomology, del, diffeomorphism, equiprobable, excenter, hyperspace, incenter, math, matroid, metadata, numerology, pseudoperfect, quasianalytic, repunit, septagon, subdiagonal, superharmonic*. By their unnatural ugliness and comical pomposity, such words betray themselves to the reader, be his intelligence ever so little.

The denominations *censor, precisian, prescriptivist,* and *sciolist* are names applied to people who make strict rules about what usage is right or wrong, and these names are not meant to be complimentary. Rules, alas, are necessary for the general public, just as are etiquette and protocol. As protocol keeps in their place people who do not know their place. so the proper use of words and grammar protects us from wasting time trying to figure out what someone is saying.

> Speaking and writing, clearly, correctly, and with ease and grace, are certainly to be acquired, by reading the best authors with care, and by attention to the best living models. (Lord Chesterfield, *Letters Written by the Late Right Honourable Philip Dormer Stanhope, Earl of Chesterfield, to His Son, Philip Stanhope, Esq., Late Envoy Extraordinary at the Court of Dresden*, J. Dodsley, London, 1774, vol. I, p. 198, Letter LXXXI)

When a new word is coined, even incorrectly, like *prequel* or *proactive*, or in a mongrel manner, like *neuroscience*, it may become established if it fills a need felt by people who do not have the fund of knowledge required to coin the correct term according to scientific principles. Similarly, usages such as unnecessarily splitting infinitives or using politically correct terminology become established practices that it is considered old-fashioned or offensive to criticize. The proliferation of such developments is irritating to people of culture and leads to the deterioration of the language; we no longer receive as much information per word as formerly, and our ears are assaulted with the most ugly concoctions and constructions. The only solution is education. The purpose of education is twofold, both positive and negative. The positive purpose of education is to present the best that the human experience has to offer in order to enable people to enjoy life and be productive members of society. The related, negative purpose of education is to prevent the freefall of language by holding the line against ill-conceived and incorrect usages. Since teachers are required to know the real meaning of what they expound, I have written this dictionary to describe the current vocabulary of our subject. It is not my intention to exhibit the behavior criticized by Voltaire, namely, to be one of those who are

> ...animé...par cette inflexibilité d'esprit que donne d'ordinaire l'étude opiniâtre des sciences de calcul. (Voltaire, *Lettres philosophiques, ou Lettres anglaises*, Éditions Garnier Frères, Paris, 1964, vingt-quatrième lettre, pp. 134–135)

> ...fired...by that inflexibility of mind which is generally found in those who devote themselves to that pertinacious study, the mathematics. (Translation found in the Harvard Classics, Easton Press Millennium Edition, 1994, vol. 34, *French and English Philosophers,* p. 158)

Instead, I find the advice given by Dean Alford in two concluding paragraphs of *The Queen's English* (pp. 278-280, 281-282) to be still sound:

> §380. But it is time that this little volume drew to an end. And if I must conclude it with some advice to my readers, it shall be that which may be inferred from these examples, and from the way in which I have been dealing with them. Be simple, be unaffected, be honest in your speaking and writing. Never use a long word when a short one will do. Call a spade a spade, not a well-known oblong instrument of manual industry; let home be home, not a residence; a place a place, not a locality; and so of the rest. When a short word will do, you always lose by using a long one. You lose in clearness; you lose in honest expression of

your meaning; and, in the estimation of all men who are qualified to judge, you lose in reputation for ability. The only true way to shine, even in this false world, is to be modest and unassuming. Falsehood may be a very thick crust, but in the course of time, truth will find a place to break through. Elegance of language may not be in the power of all of us; but simplicity and straightforwardness are. Write much as you would speak; speak as you think. If with your inferiors, speak no coarser than usual; if with your superiors, no finer. Be what you say; and, within the rules of prudence, say what you are.

§386. These stray notes on spelling and speaking have been written more as contributions to discussion, than as attempts to decide in doubtful cases. The decision of matters such as those which I have treated is not made by one man or set of men; cannot be brought about by strong writing, or vehement assertion; but depends on influences wider than any one man's view, and takes longer to operate than the life of any one generation. It depends on the direction and deviations of the currents of a nation's thoughts, and the influence exercised on words by events beyond man's control. Grammarians and rhetoricians may set bounds to language: but usage will break over in spite of them. And I have ventured to think that he may do some service who, instead of standing and protesting where this has been the case, observes, and points out to others, the existing phenomena, and the probable account to be given of them.

Finally, I express my gratitude to editor Maria E. denBoer and my student Reuben Bernstein-Goff, without whose help I would have failed to submit the final manuscript in the form desired by the Press.

Origins of Mathematical Words

A

a-, an-, in-, im-, un- Each of these is a prefix that negates the word to which it is appended. However, these prefixes are not to be used interchangeably; *alpha privativum* (ἀ- or ἀν-), that is, *a-* or *an-* (before a vowel), is placed before Greek words, *in-* (or *im-* before a labial consonant, that is, before *b* or *p*) before Latin words, and *un-* before Germanic words. When this rule is violated, as, for example, in the case of the internet lingo *unsubscribe,* the resulting word is low, although in cases like *unequal,* the construction must be accepted due to immemorial custom. *Verbs* of Latin origin are made negative by appending the prefix *dis-,* for example, *disassociate, discredit.* The Latin prefix *in-* or *im-* serves double duty as the preposition *in,* which means *in* or *into,* so special care is necessary when compounding words with it.

The Latin adverb *non* means *not.* It may be considered naturalized English and should be used to negate words for which the addition of *a-, an-, in-, im-,* or *un-* is unprecedented. Thus, we say *non-compact,* not *incompact.*

The choice of the wrong prefix will lead to confusion. One morning (November 3, 2011) viewers heard the word *amoral* used incorrectly on an episode of *Judge Judy,* as if it meant *immoral.* The culprit added a Greek suffix to a Latin adjective. People who know neither Greek nor Latin know the word *moral,* and some of them even know that the prefix *a-* negates the sense of the following adjective to which it is attached. The result is the word *amoral* intended to mean *immoral.* However, if *amoral* is to mean anything, it must mean *pertaining to love,* from the Latin *amor, love,* and this is the only meaning it has for people who know something. Even in cases where there is no confusion as to the intent of the author, the choice of the wrong prefix will at least lead to awkwardness. For example,

1

the cover of the June/July 2012 issue of the *Notices of the American Mathematical Society* announces an article within entitled "Incomputability after Alan Turing," but there is no such word as *incomputability*, though it is formed correctly after classical models. The word does not exist except as an error because it has never been used by polished authors. The correct word is *non-computability*. This illustrates the precariousness of forming new words according to rule from foreign languages without reference to the usage of the first class of writers. In this regard, not entirely irrelevant is the comment of Thomas Paine:

> The best Greek linguist that now exists does not understand Greek so well as a Grecian plowman did, or a Grecian milkmaid; and the same for the Latin, compared with a plowman or a milkmaid of the Romans.... (Thomas Paine, *Age of Reason*, The World's Popular Classics, Books, Inc. Publishers, New York and Boston, no date, p. 48)

The fact that a word appears in the *Oxford English Dictionary* does not imply that it is a good word; you will find *incommutative* in that lexicon, but though a cautious fellow may call it *rare*, a frank one will call it *wrong*. The sanction of existence can only be imparted to a word through its use by a polished author, like, for example, Lord Chesterfield, in whose *Letters to His Son* we may find the following instructive passage:

LETTER CXXXII

London, September the 27th, O. S. 1748.

DEAR BOY,

I have received your Latin Lecture upon War, which, though it is not exactly the same Latin that Cesar, Cicero, Horace, Virgil, and Ovid spoke, is, however, as good Latin as the *erudite Germans* speak or write. I have always observed, that the most learned people, that is those who have read the most Latin, write the worst; and that distinguishes the Latin of a Gentleman scholar, from that of a Pedant. A Gentleman has, probably, read no other Latin but that of the Augustine age; and therefore can write no other: whereas the Pedant has read much more bad

Latin than good; and consequently writes so too. He looks upon the best classical books, as books for school-boys, and consequently below him; but pores over fragments of obscure authors, treasures up the obsolete words which he meets with there, and uses them, upon all occasions, to show his reading, at the expence of his judgment. Plautus is his favourite author, not for the sake of the wit and the *vis comica* of his comedies; but upon account of the many obsolete words, which are to be met with no where else. He would rather use *olli* than *illi*, *optumè* than *optimè*, and any bad word, rather than any good one, provided he can but prove, that, strictly speaking, it is Latin; that is, that it was written by a Roman. By this rule, I might now write to you in the language of Chaucer or Spenser, and assert that I wrote English, because it was English in their days; but I should be a most affected puppy if I did so, and you would not understand three words of my letter. All these, and such-like affected peculiarities, are the characteristics of learned coxcombs and pedants, and are carefully avoided by all men of sense.

I dipped, accidentally, the other day, into Pitiscus's preface to his Lexicon; where I found a word that puzzled me, and that I did not remember ever to have met with before. It is the adverb *praefiscinè*; which means, *in a good hour*; an expression, which, by the superstition of it, appears to be low and vulgar. I looked for it; and at last I found, that it is once or twice made use of in Plautus; upon the strength of which, this learned pedant thrusts it into his preface.... (Lord Chesterfield, *Letters Written by the Late Right Honourable Philip Dormer Stanhope, Earl of Chesterfield, to His Son, Philip Stanhope, Esq., Late Envoy Extraordinary at the Court of Dresden*, J. Dodsley, London, 1774, vol. I, pp. 341–342)

abacus The *abacus* is a frame with rods and beads used for computation. It is derived from the Latin word *abacus*, which is itself the Latinization of the Greek word ἄβαξ, which means *a square board*. Its use to mean *a counting board* is found in the satirical poet A. Persius Flaccus (A.D. 34–62) at line 131 of Satire 1:

Nec qui abaco numeros et secto in pulvere metas
Scit risisse vafer, multum gaudere paratus,
Si cynico barbam petulans nonaria vellat.

> ...nor the fool who ridicules arithmeticians at the abacus and
> solid geometers with their cones, the sort of fellow who is
> amused when a whore pulls the beard of a grave philosopher....

It is incorrect, though tempting, to derive the word from the names of the first three letters of the Latin alphabet, *a-ba-ca*, but in this case one would expect the word to be *abaca, -ae*, not *abacus, -i*.

Abelian Not to capitalize the first *a* is a mistake. Adjectives formed from proper names must always be capitalized, for they otherwise look ridiculous. The capitalization alerts the general reader (for the expert would already be aware) to the origin of the word, that it is taken from a person's name. Humanity defers in such matters to those with taste, and only those with no taste write *abelian*. The adjective is from the name of Nils Abel (1802–1829), and the term is a synonym of *commutative* and is applied to groups. Abel was not so conceited as to use the term himself; it was invented by Camille Jordan (1838–1922). It is not to be confounded with an earlier use of the word, as the name of a type of African heretics who abstained from sex, even with their wives, in imitation, so they imagined, of the holy Abel, son of Adam and Eve. The ending *-ian* in *Abelian* and *Eulerian* is from the Latin adjectival suffix *-anus* added to the stem of the names after the connecting vowel *-i-* was inserted. The names in Latin are *Abelus* and *Eulerus*, not *i*-stems like *Abelius* or *Eulerius*; the insertion of the letter *i* was for ease of pronunciation, as in the case of *Christian*.

-able This is the degeneration of the Latin suffix *-abilis, -e*, which, when added to a word, imparts the notion of capability. Thus, *legible* means *capable of being read*. When imposed on a word not of Latin origin and then turned into a noun, the concoction may sound ridiculous, such as *reachability*, which is a property of an ordered pair of vertices of a graph defined by Finkbeiner and Lindstrom on page 217 of *A Primer of Discrete Mathematics*, W. H. Freeman and Company, New York, 1987. The suffix *-abilis* is from the adjective *habilis, -e*, which means *fit, apt*, derived from the verb *habeo, habere, habui, habitus*, which means *to have*.

abscissa, ordinate The line segment in the plane from the origin to the point $(x,0)$ is called the *abscissa*, that is, the segment *cut off* from the horizontal axis, *abscissa ab axe*. The segment from $(x,0)$ to (x,y) is then called the *ordinate* of (x,y). By transumption, we are allowed to say that x is the abscissa, and y the ordinate, of the point (x,y). *Abscissa* is the past participle, feminine, of the verb *abscindo, abscindere, abscidi, abscissus, to lop off,* and modifies *linea, line,* understood. The word was first used in English in 1698 by Abraham De Moivre, *Philosophical Transactions* XX, page 192. Leibniz had made it popular and had used it, for example, in a paper written in Latin in the *Acta Eruditorum* 3 (1684), pages 467ff. He is similarly responsible for the general usage of the term *ordinate*. *Ordinate* comes from the past participle *ordinatus* of the verb *ordino, to set in order*. The suffix *-ate* in English is a sign that a word is taken from the past participle of a Latin verb of the first conjugation. Information on the first use of technical terms can be reliably found in the *Oxford English Dictionary* and Struik's *Concise History of Mathematics*.

absolute value The *absolute value* of a real number is its magnitude irrespective of its sign. The Latin word for the same idea is *modulus*, which has survived in the language of complex analysis. The adjective *absolute* is from the Latin past participle *absolutus, -a, -um*, the fourth principal part of *absolvo, absolvere, absolvi, absolutus, to release or free* from the tyranny of its sign or direction. The term appears in *A Textbook of Analytic Geometry* by James Mill Peirce (1857); it had earlier been used by Carnot in *Mémoire sur la relation qui existe entre les distances respectives de cinq points quelconques pris dans l'espace*, page 105. Weierstraß introduced the standard notation $|x|$ in 1841.

absorbing The English verb *absorb* is the stem of the Latin verb *absorbeo, absorbere, absorbui, absorptus*, which means *to swallow or gulp down*. The terminology *absorbing states* is sometimes used in probability theory for what are more commonly known as invariant sets.

absorption The fourth principal part *absorptus, -a, -um* of the Latin verb *absorbeo, absorbere, absorbui* is rare. From it is formed the noun *absorptio, absorptionis* meaning originally *a drink* or *a beverage*.

abstract This is an adjective formed from the past participle *abstractus, -a, -um* of the Latin verb *abstraho, abstrahere, abstraxi, abstractus*, which means *to drag away or remove* from something. It acquired the colloquial meaning of r*emoved from experience* and therefore *difficult to understand* among those not accustomed to consecutive thought. *Abstract algebra* is the study of groups, rings, and fields, found difficult by the multitude, whereas plain *algebra* would be understood to be the study and application of the elementary laws of arithmetic. The names *abstract algebra, modern algebra,* and *higher algebra* for the study of the structures mentioned above derive from the titles of famous books on the subject by authors like Albert, Birkhoff and MacLane, Herstein, Jacobson, and Van der Waerden.

abundant number A natural number that is less than the sum of its proper positive integral divisors is called *abundant*, from the Latin present participle *abundans, abundantis* of the verb *abundo, abundare, abundavi, abundatus*, which means *to overflow*; the preposition *ab* means *from*, and the noun *unda* means *wave*. The idea is Greek, as Nicomachus (*circa* 80–120) speaks of ἀριθμοι ὑπερτελεῖς or "beyond-perfect numbers" in his *Introduction to Arithmetic*, section 14. The Roman translators preferred *abundant* to *transperfect*. There are infinitely many abundant numbers since every positive integer of the form $3(2^k)$, $k = 2, 3, 4,\ldots$ is abundant. Natural numbers that equal the sum of their proper divisors are called *perfect (q.v.)*; those that are greater than that sum are *deficient (q.v.)*. A recent well-written paper on the subject is by Roger Webster and Gareth Williams, "Friends in High Places," *Mathematical Spectrum*, vol. 42, no. 2 (2010), pages 54–58.

academy This name for a school is derived from the Greek proper name Ἀκάδημος, who was the fellow who allowed Plato to teach on his property. *Inter silvas Academi quaerere verum.* The word became the name for a first-class school and for the community of scholars. Alas, the eminence of this high word has been insulted by the modern phenomenon of highfalutin names for nonsense activities. Recently I learned that people who want to be bartenders learn the trade at *bartending academies*. Hairdressers are now instructed in *academies of cosmetology*. One of my mathematics advisees was shipped to Germany for "study away" and, once off the plane, attended a meeting of the *innovation academy* in Freiburg im Breisgau. A terrible recent abuse of the majesty of the related adjective *academic* was recently brought to my attention. This adjective formerly indicated a connection with scientific learning, for example, the *academic philosophy* of Hume. I am told that the adjective is used today in American high schools to describe the course of studies of students who are not qualified to take "advanced placement" courses. Thus, we are moving toward the day when to say that a student follows an *academic* program will mean that he is considered an underachiever.

acceleration From the Latin verb *accelero*, which means *to quicken*, itself derived from the composition of the prefix *ad-*, which intensifies the effect of the following adjective *celer*, which means *swift*. The *acceleration vector* is the derivative of the velocity vector with respect to time.

accumulation point The Latin noun *cumulus* means *a heap*, and from it and the preposition *ad* (*to*) the verb *accumulo* (*ad* + *cumulo*), *accumulare, accumulavi, accumulatus* is formed, meaning *to heap up*. From the fourth principal part comes the noun *accumulatio, -onis* with the meaning *a heaping up*. An *accumulation point* of a sequence is a point, every open interval around which contains an infinite number of points of the sequence.

accuracy *Accuratio* is a Latin noun meaning *carefulness* composed of the prefix *ad-* (changed to *ac-* by assimilation to the following *c*) and the noun *cura*, which means *care*.

acnode The Latin adjective *acer, acris, acre* means *sharp*. The *r* is not part of the root, which is *ac-*, as in *acuo, acuere, acui, acutus*, which means *to be sharp*, in *acus, -ūs*, which means *a needle*, and in *acies*, which means *keenness, edge, line of battle*. The noun *node* comes from *nodus*, the Latin word for *knot*. An *acnode* is a cusp formed at a point of a continuous curve that is linear on one side of that point but not linear on the other. According to the *Oxford English Dictionary*, the word may be traced at least as far back as George Salmon, *Higher Plane Curves*, second edition, page 23, a work published in 1873.

actuary This is a fellow who predicts when someone will die and the financial consequences thereof. Actuaries are well paid employees of insurance companies with a solid foundation in probability, statistics, and economics. According to the *Oxford English Dictionary*, the earliest use of the word in this mathematical sense is in Macaulay's *History of England*, vol. I, page 283 (1849). The word is the English form of the Latin noun *actuarius* and was the name for someone in charge of official records, the *actus*, just as the *ostiary*, from the Latin *ostiarius*, is the man in charge of the *ostium*, or church door.

acute This adjective comes from the Latin *acutus, -a, -um*, the fourth principal part of the verb *acuo, acuere, acui, acutus*, which means *to be sharp*. It was the literal translation of the Greek word used by Euclid, ὀξύς. It appears in mathematics for the first time in the Latin version of the *Elements* by Boëthius (late fifth or early sixth century A.D.).

acyclic This adjective means *not cyclic*, and is the combination of *alpha privativum* and the adjective *cyclic*, *q.v.*

add The Latin word for *to add* is *addo, addere, addidi, additus*; the English word is obtained by removing the infinitive ending *-ere*.

addition The late Latin noun *additio* is from the classical verb *addo*, which means *to give to*.

additive The late Latin adjective *additivus, -a, -um*, which means *added, annexed*, was derived from adding the adjectival ending *-ivus* to the stem of the fourth principal part *additus* of the verb *addo*.

additivity This is a modern mathematical term formed as if there were a Latin noun *additivitas*, meaning *the condition of being added or annexed*, from the adjective *additivus*.

ad infinitum This is a Latin prepositional phrase meaning *to infinity* or *and so on*, the mathematical equivalent of *in saecula saeculorum, unto centuries of centuries*, and the Hebrew לְעוֹלָם, *unto the indefinite future*.

adjacent This comes from the Latin present participle *adiacens, adiacentis* of the verb *adiaceo, adiacere*, which means *to lie next to*. The letter *j* is a form of *i* used in late Latin when the letter *i* would fall before another vowel. It was also used at the end of a word, usually but not necessarily after a previous *i*; for example, the name *Basilii* might be written *Basilij*.

adjoined This word is derived from the Latin adjective *adiunctus*, the fourth principal part of the verb *adiungo, adiungere, adiunxi, adiunctus*, which means *to join to*. The verb is the composition of the prefix *ad-* (*to*) and the verb *iungo* (*to join*).

adjoint This noun has the same root, *adiunctus*, as the adjective *adjoined*. It is the name of a person taken on to assist someone else. The *adjoint* A' of a linear transformation A on a vector space Υ is an associated transformation on the dual space Υ' defined by

$$[A'(f)](x) = f[A(x)] \text{ for all } x \in \Upsilon.$$

adjunction The Latin verb *iungo, iungere, iunxi, iunctus* means *to join*, and the preposition *ad* means *to*. The combination produces the verb *adiungo, adiungere, adiunxi, adiunctus* with the meaning *to join to*. From the fourth principal part comes the noun *adiunctio, -onis* with the meaning *a joining to*, from whose stem comes the English word. If F and M are fields, $F \subset M$, $a \in M$, and $a \notin F$, the *adjunction* of an element a to the field F is the intersection of all subfields of M containing both a and F.

affine This adjective is derived from the Latin word *adfinis*, which means *neighboring*, which is itself the composition of the intensifying prefix *af-* (changed on account of assimilation from *ad-*) and the

9

noun *finis*, *boundary*. The use of the Latin adjective *affinis* in mathematics can be traced at least as far back as Euler.

aggregate This is a synonym for *set*. See the following entry.

aggregation *Grex* means a *herd* or *flock* in Latin; *ag-* is the prefix *ad-* (*to*) with the *d* changed to *g* for the sake of euphony. Thus, *aggregatio, -onis* is *an addition to a herd or flock*. In algebraic expressions, pairs of parentheses, braces, and brackets are *symbols of aggregation*.

aleph This is ℵ, the first letter of the Hebrew alphabet. It was admitted into mathematics by Cantor, who used $\aleph_0$, *aleph-null*, for the cardinal number of the set of positive integers. The use of a Hebrew letter was daring since such unfamiliar symbols tend to be copied incorrectly over time and become unseemly. The practice is not to be recommended. In 1815 the founder of Allegheny College, Meadville, Pennsylvania, chose a Hebrew verse Isaiah L 1 for the motto of his school, and one hundred years later twelve of the eighteen letters had been transformed into other letters or into illegible squiggles by generations of uncomprehending copyists.

algebra The scholar al-Khowarizmi wrote in the first half of the ninth century حساب الجبر والمقابلة, *The Book of Restoration* (adding a negative term to the other side of an equation) *and of Coming Together* (adding like terms on one or both sides of an equation), which title was translated into Latin by Robert of Chester in the twelfth century as *Liber algebrae et almucabala*; he knew no Latin equivalents for two of the nouns, so he just transliterated them. *Algebra* is the corruption of the Arabic word الجبر, which means *restoration, the manipulations whereby a broken bone is reset*. The prefix *al-* at the beginning of a word is a sign that the word is of Arabic origin, for it is the Arabic definite article *the*. The corruptions of Arabic words are a prominent part of the vocabulary of mathematics, for the rebirth of our subject after the period of the "Dark Ages" occurred when the caliph of the Muslims obtained a manuscript of Euclid from the Byzantine emperor. Here is the story, taken from Lo Bello, *The Commentary of al-Nayrizi on Book I* *of Euclid's Elements of Geometry*, Brill Academic Publishers, Inc., Boston and Leiden, 2003, page 23.

> In his *Muqaddimah*, or introduction to history, the statesman, jurist, historian, and scholar Ibn Khaldun (1332–1406) related how Greek learning came to the attention of the Arabs after they took Syria from the Eastern Roman Empire and settled there:
> "Then they desired to study the philosophical disciplines. They had heard some mention of them by the bishops and priests among their Christian subjects, and man's ability to think has aspirations in the direction of the intellectual sciences. Abu Jafar al-Mansur [754–775], therefore, sent to the Byzantine Emperor [Constantine V Copronymus] and asked him to send him translations of mathematical works. The Emperor sent him Euclid's book and some works on physics. The Muslims read them and studied their contents. Their desire to obtain the rest of them grew. Later on, al-Mamun [813–833] came. He had some [scientific knowledge]. Therefore, he had a desire for science. His desire aroused him to action in behalf of the intellectual sciences. He sent ambassadors to the Byzantine Emperors [Leo V, Michael II, Theophilus]; they were to discover the Greek sciences and have them copied in Arabic writing; he sent translators for that purpose. As a result, a good deal of the material was preserved and collected." (Ibn Khaldun, *The Muqaddimah, an Introduction to History*, translated from Arabic by Franz Rosenthal, edited by N. J. Dawood, Princeton University Press, 1989, p. 374)

As for the corruptions which mathematical words suffered upon transliteration from Greek to Arabic, and from Arabic to Latin, we may note the following causes.

> The names of the ancient mathematicians were transfigured, often beyond recognition, as they passed from Greek to Arabic or from Arabic to Latin. Those that made both passages were especially corrupted. The ancient and medieval personalities lacked the critical sense to transliterate the names scientifically; they did not follow the principle that everyone is entitled to his name, and that one's tongue would not break if the name were pronounced as it ought. As a result, we are faced with such deplorable and often unintelligible concoctions as Sanbeliqiyus, Sambelichius, Assamites, Aghanis, Irun, Irinus, Herundes,

Deurus, Hermydes, and Banbus. Readers of the Bible, of course, will not be surprised by these transformations, for they will know that such transformations as Jesus and Isaac, for example, are the products of similar mutilation.

The following are the chief causes, other than human carelessness, for the barbarization of the names of the mathematicians:

1) The Greek letters π (*pi*) and ψ (*psi*) have no equivalent in Arabic. The Arabs therefore approximated π, for example, with their letter for *b*. This is the reason for the *b* in Sanbeliqiyus (Simplicius).

2) The Arabic alphabet consists of consonants only, and although there are marks to indicate the vowel sounds, they are almost never written, except in editions of the Quran or in books for children. This meant that oral tradition, rather than the written word, determined the vowels to be used, and this tradition often failed. Thus we have the *a* in Sanbeliqiyus. Vowel sounds are notoriously changeable in the development of languages.

3) The Arabic letters for *b, n, t, y,* and *th* cannot be distinguished from one another when they appear at the beginning or in the middle of a word except by placing one, two, or three dots (points) above or below them, but these dots, like the marks for the vowels, were usually omitted. This fact, together with that noted in 1) above, accounts for how *Pappus* became *Banbus*. In "unpointed" texts, there are similar difficulties with six other pairs of letters, for example, between *r* and *z,* and between *s* and *sh.*

4) The Arabs inserted vowels to prevent three consonants from coming together. This accounts for the *e* in Sanbeliqiyus.

5) The Greeks did not always write the rough breathing (initial *h*), so the Arabs read *Heron* as *Eron,* whence they produced *Irun.*

6) The Arabic consonants *y* and *w* were also used to indicate the long vowels *ī* and *ū* (or *ō*), respectively, but this was not always done, and so, as a result, the *ō* in *Heron* was lost and ended up as the second *i* in *Irinus,* the Latinization of *Irun.*

7) The *l* of the Arabic definite article *al* is assimilated in pronunciation to certain following consonants, among which is *n*; *al-Nayrizi*, therefore, is pronounced *an-Nayrizi*, and this accounts for the lack of an *l* in *Anaritius*, the medieval Latin equivalent for *al-Nayrizi*.

8) The medieval Western authors frequently Latinized the names they received by adding the ending *-us* of the masculine nouns of the second declension; this accounts for the *-us* at the end of *Anaritius* and *Irinus*.

9) There are many letters in Arabic that have no equivalent in Latin. In particular, the Romans used *z* only when they were transcribing Greek words. Thus, the *z* in *al-Nayrizi*, though occasionally preserved in the form *Anarizius*, almost always becomes a *c* or a *t*, so as to produce the usual form of the name in the Latin West, *Anaritius*.

10) How did the *m* of *Simplicius* become the *n* of *Sanbelichius*? The metamorphoses of these two letters have long agitated philologists. "That *n* and *m* readily interchange is known to us" (William Wright, *Lectures on the Comparative Grammar of the Semitic Languages*, ed. W. Robertson Smith, Philo Press, Amsterdam, 1966, p. 144). Euphony, the demand for an ease of pronunciation aggravated by the presence of the following labial, may be accounted responsible for the aforementioned transformations. (Anthony Lo Bello, *The Commentary of al-Nayrizi on Book I of Euclid's Elements of Geometry*, Brill Academic Publishers, Inc., Boston and Leiden, 2003, pp. 18–20)

algebraic This is an example of a good macaronic word. It is macaronic because the Greek adjectival suffix -ικός has been added to an Arabic stem; it is good because there is no alternative, the addition of the corresponding Arabic suffix being beyond what can be expected of Western word makers. A bad macaronic word is *untypical*, which consists of Germanic (*un-*), Greek (*-typic-*), and Latin (*-al*) elements; *typic* would have been serviceable as an English adjective, yet the stem of the Latin adjectival ending *-alis* was nevertheless superfluously added on at the end. Furthermore, ἀ- was the proper prefix to negate the adjective *typic*. The algebraic numbers are those real numbers that are roots of polynomials with integer coefficients.

algorithm This is the corruption of the Arabic name الخوارزمي, which means *the fellow from the town Khowarizm*. See the entry **algebra** above.

aliquot The Latin indeclinable adjective *aliquot* means *so many*. The *aliquot parts* of a number is a term defined by Euclid in the *Elements*, Book V, Definition 1, to be the proper and non-unit positive integral divisors of a natural number. Thus, the aliquot parts of 8 are 2 and 4. Euclid's term was μέρος (plural τὰ μέρη), *part*, which here means *submultiple*.

-alis, -ale This is a Latin adjectival suffix added to the stem of nouns to make the corresponding adjectives with the meaning of *pertaining to*. Thus to the stem *ordin-* of the noun *ordo, ordinis* one produces the adjective *ordinalis*. If the stem contains an *l*, one adds *-aris* instead of *-alis* to avoid cacophony. Thus to the stem of *singuli* one adds *-aris* to produce the adjective *singularis*. The suffixes *-elis*, *-ilis*, and *-ulis* are also occasionally found with this force. In creating new words, this suffix should not be added to a stem that is not Latin.

Almajest, the This is the Arabic المجستي, part translation and part transliteration of the Greek ἡ μεγίστη; the Greek definite article is translated, and the adjective is transliterated. The title of the masterpiece of Ptolemy was ἡ μεγίστη σύνταξις, which means *The Greatest Arrangement*.

alphabet This word for the set of letters of the Greek alphabet is the concatenation of the first two Greek letters, *alpha* and *beta*. According to Weekley, the word is first found in English in the sixteenth century.

alternating The Latin adjective *alter, altera, alterum* means *one of two*, and from it the adjective *alternus, -a, -um* is derived, meaning *one after the other*. There then proceeds the verb *alterno, alternare, alternavi, alternatus* with the meaning *to come one after the other*, and from its fourth principal part is derived the English verb *to alternate*. The *alternating* *group* A_n is the subgroup of the symmetric group S_n consisting of all even permutations of the set $\{1,2,3,\ldots,n\}$.

altitude This is the Latin noun *altitudo*, which means *height*. The adjective *altus, -a, -um* means *tall* or *deep*.

alysoid This is a mistake for the Greek name of the curve more commonly known by its Latin name, the *catenary*, *q.v.*, and it would be an affectation to use it nowadays. The correct form would be *halysoid*. Whoever concocted it ignored the rough breathing on the initial *alpha*. The word *alysoid* means *the shape* (οἶδος) *assumed by a chain* (ἅλυσις). Breathings and subscripts in Greek words are often ignored by the unlearned, who imagine that those tiny marks are mere specks of dirt.

ambiguous *Amb-, ambi-, am-, an-* are inseparable Latin prefixes related to the Greek preposition ἀμφί and are added to words to indicate *on both sides, around, round, about*. *Ambigo* (*ambi* + *ago*) is a third-declension verb meaning *to go about or around, to doubt, to hesitate*. *Ambiguitas* is a good classical Latin word meaning *ambiguity*, and *ambiguus* is the corresponding Latin adjective meaning *uncertain, doubtful*. The Latin adjectival ending *-us* of the first- and second-declension masculine adjective was regularly changed in English to *-ous*. This should not be done in the case of Latin or Greek proper nouns; for example, the Greek name Κύριλλος (Cyril) should not be written *Kyrillous*.

American spelling The name *America* is derived from the proper name Ἀμαλάριχος or *Amaláricus*, meaning *tireless ruler*, which was common among the Ostrogoths; it came into Italian as *Amerígo* and was the first name of the explorer Vespucci, from whence it was adopted as the name of our continent. The accent on the antepenult is due to the Spanish pronunciation *Américo*. *American spelling* is a term used to describe the changes introduced by Noah Webster, who modified the received system in accordance, as he imagined, with the demands of reason. The topic is discussed fully in the article "American Spelling" by Herbert Thurston, S.J., in *The Nineteenth*

Century, vol. 60, no. 356 (October 1906), pages 606–617. This system of American spelling is established in the United States, and in accordance with the rule *Roma locuta, causa finita*, it should be employed by American authors. It was, however, not a good idea to begin with.

One of the crimes Webster committed was to change the spelling of certain English words when he published his dictionary. For example, the English spelled *author a-u-t-h-o-r* but *honor h-o-n-o-u-r*. That is to say, the English inserted a *u* in some Latin nouns that ended in *-or*, but not in others. Webster removed all the *u*'s wherever they occurred in such words, and we follow his example to this day in America. In England, of course, Webster had no authority, and his dictionary counted for nothing. Now we may ask, why do the English spell some of the Latin nouns, like *author*, with *-or* and others like *honour* with *-our*? The reason is that the Latin language came into England in two waves. The first began with Julius Caesar's invasion of 55 B.C. and continued through the following centuries when Latin-speaking missionaries brought the Catholic religion to the island; the second began with the Norman invasion of 1066. However, the Latin that came in with William the Conqueror came in with a French dress, and its spelling was affected by the French language of the time.

> Twas Greek at first; that Greek was Latin made:
> That Latin, French; that French to English straid.
> (Dr. Richard Farmer, 1735–1797, *Essay on the Learning of Shakespeare*, 1767, quoted by Alford, p. 21)

Thus, the word *author* came into common use after the first wave sometime in the first century A.D.; it is therefore spelled exactly as in Latin, without the *u*. The word *honour*, however, came into common use only after the Norman invasion of the eleventh century, and the Normans spelled it *honneur*. Thus, if you learn the English spelling, you know when the word came into the language by noticing whether it has the *u* or not. If you learn the American spelling, you cannot tell. Thus Americans have been deprived of knowledge by an educational system that uses Webster's spelling.

In the preceding comment, we considered a change introduced by a single man that was freely adopted by a whole country. It sometimes happens that *governments* make changes in the language of the land, and this has always been for the worse. The reason is that these changes are meant to make learning easier for slow learners and only end up by erasing knowledge. One of the first changes made by the Bolshevik government in Russia was the abolition of several letters in 1918. There were two results. First, no one who learns the modern system can study etymology easily since the abolished letters had a reason for being there, and second, young people cannot read books published before 1918. The following comment on the matter by a member of the imperial family is not without merit:

> One Morning, on opening the abominable Bolshevist newspaper, so difficult to read on account of the new orthography made by illiterates, I saw with a feeling of stupor a decree... (Princess Paley, *Memories of Russia 1916–1919*, Herbert Jenkins Limited, London, 1924, p. 251)

Similarly, *Sütterlin* script and *Fraktur* type are no longer a required part of the curriculum in Germany. The result is that modern Germans cannot read any letters written in *Sütterlin* or any books written in *Fraktur*. One language that cannot change by a decree of the government is Arabic since (the Muslims believe) it is the language chosen by the Deity to communicate his revelation to men. Therefore it cannot be touched. Suggestions made from time to time to allow phonetic spelling in English should be opposed since the history of the words would thereby be erased, and the slow learners would not learn how to spell anyway. What phonetic spelling would lead to may be deduced from the result of that license that allows Americans to spell the names of their children phonetically (as they imagine).

amicable This is the Latin adjective *amicabilis*, which means *friendly*. In English it should be pronounced *a-mi´-ca-ble* not *a´-mi-ca-ble*, for so says Dr. Johnson. It is used of a pair m, n of positive integers such that m is the sum of the proper divisors of n, and n is the sum of the proper divisors of *m*. The numbers *m* and *n* are also called *friendly numbers*. Euclid was the first to study friendly numbers, which he denominated ἀριθμοὶ φίλιοι. He knew of but one such pair, *viz.*, (220; 284), which was probably handed down by the Pythagoreans. This pair and the next four pairs to be found are exhibited in the following table:

Pythagoras (*circa* 500 B.C.)	220	284
Fermat (1636)	17,296	18,416
Descartes (1638)	9,363,584	9,437,056
Euler (1747)	2,620	2,924
Paganini (1866)	1,184	2,924

Others have since been discovered by computers, which are capable of factoring astronomically large integers. This is an unsatisfactory situation since the verification of the calculations depends on a machine, which is assumed to be infallible. It is a matter of faith that the results thereby obtained are true. As I write, the Wikipedia entry tells me that in 2007 there were twelve million known pairs of amicable numbers. I do not know whether to believe it.

amplitude This is from the Latin noun *amplitudo*, which means *fullness, width*; it is derived from the adjective *amplus, -a, -um*, which means *wide*. It is the name of half the distance between the maximum and minimum values of a curve describing simple harmonic motion.

analog It is incorrect to drop the *-ue* from the French ending. Words like *analog* and *catalog* are a concession to ignorance. See the entry **American spelling**.

analogy This is the Greek ἀναλογία, which means *proportion*, from the verb ἀναλογίζομαι, *to count up, think over, calculate*. The preposition ἀνά means *up*, and the verb λογίζομαι means *to count or reckon*, from λόγος, one of whose meanings is *calculation, reckoning*.

analysis A *setting free* (λύσις) by proceeding *upwards* or *backwards* (ἀνά) from a point. Originally it was the equivalent of what we would call today a proof by reversible steps, starting with the conclusion desired and going back by "if and only if" statements to the hypothesis. It is the modern name for that branch of higher mathematics that developed out of calculus.

analysis situs This means the *analysis of position*, and is another name for *topology*. Poincaré defined it as "a branch of Geometry...which describes the relative positions of lines and surfaces without regard to their size." (This translation is to be found in Ronald Calinger, *Classics of Mathematics*, Prentice-Hall, 1995, p. 754.) The phrase is awkward since it consists of a Greek nominative followed by a Latin genitive. Nevertheless, it has the *imprimatur* of Leibniz, and is therefore beyond criticism.

analytic This is the Greek word ἀναλυτικός formed from the noun ἀνάλυσις, which means *a loosening, releasing, dissolution*. The verb is ἀναλύω, which means *to unloose, set free, do away with*.

angle This is the Latin word *angulus*, which was the translation used by Boëthius for the Greek γωνία in his edition of Euclid's *Elements*. It is related to the Greek adjective ἀγκύλος, *crooked, bent*, from the noun τὸ ἄγκος, *the bend or hollow, particularly of the arm*.

annihilator This is a noun of agent formed from a Latin verb *annihilo, annihilare, annihilavi, annihilatus* created by St. Jerome (Epistle 135) and meaning *to bring to nought*. It is a combination of the preposition *ad* (*to*) and the Latin noun *nihil*, which means *nothing*. The *annihilator* of a subspace W of a vector space V is the set of all linear functionals defined on V that map the elements of W to zero.

anomaly [of a point in the polar plane] This is another name for the *amplitude*, the angle that is one of the two polar coordinates of a point. It is the Greek word ἀνωμαλία, *irregularity*, derived from the combination of *alpha privativum* and the adjective ὁμαλός, *even*.

annuity The medieval Latin noun *annuitas* is derived from the adjective *annuus, that which lasts a year*, a word that itself proceeds from the noun *annus, year*. An *annuity* is the promise to pay a certain amount on an annual basis for life. In 1751, Euler reported in the *Memoirs of the Berlin Academy* that according to his calculations, a payment of 350 crowns should purchase a newborn Prussian baby a deferred annuity of 100 crowns to commence on the twentieth birthday and to continue for life. This implies a mean lifespan of 23.5 years in Prussia at that time. See page 242 of *A History of the Mathematical Theory of Probability from the Time of Pascal to that of Laplace* by Isaac Todhunter, M.A., F.R.S., Chelsea Publishing Company, Bronx, New York, 1965, a textually unaltered reprint of the first edition published by Cambridge University in 1865.

annulus The Latin noun *anulus* means *a ring*, especially for the finger. The addition of the second *n* is a medieval mistake due to confusion with the common noun *annus*, the Latin word for *year*. This mistake may be traced back to the Middle Ages, when the "fisherman's ring" of the Roman pontiff was called the *annulus piscatoris*. Double consonants are distinguishable in pronunciation from single consonants in Latin, so this sort of mistake is rare. In English, however, there is no difference, and as a result Noah Webster sought to abolish knowledge by cancelling the doubling of the consonant in certain cases. See Herbert Thurston, *American Spelling, passim*. The *annulus* is the plane region between two concentric circles.

ANOVA This is an acronym for the *analysis of variance*, a branch of mathematical statistics that deals with the problem of what sources are to blame for the variation in random samples. Acronyms in mathematics are to be avoided; they must not be multiplied beyond necessity. They are ugly cant and lead to confusion, as it is impossible to keep track of them all. The article "Debunking Myths about Gender and Mathematics Performance" in the January 2012 issue of *The Notices of the American Mathematical Society* (pp. 10–21) is particularly "rich" in acronyms. One finds there ED, EPO, GDP, GEI, GGI, H&S, IMO, OECD, PISA, POL, STEM, TIMSS, VR, and some others that may have been overlooked by the investigator. GGI, by the way, stands for *Gender Gap Index*, and "it is measured on a 0–1 scale, with 1.00 being complete gender equity [p. 12b]."

The proliferation of acronyms is a peculiarity of modern English usage to be found in society at large, from which influence neither mathematicians nor their colleagues on American campuses are immune.

antecedent This is from the present active participle *antecedens, antecedentis* of the verb *antecedo*, which means *to go before, to precede*. The preposition *ante* means *before*, and the verb *cedo, cedere, cessi, cessus* means *to go, to go away*.

antichain This is a strange-sounding word, the combination of the Greek prefix ἀντί- (*against, opposed to, opposite to*) and the English *chain*, which is the metamorphosis, *via* French, of the Latin *catena*. Finkbeiner and Lindstrom, in *A Primer of Discrete Mathematics*, W. H. Freeman and Company, New York, 1987, page 125, define an antichain as a completely unordered subset of a set S, that is, there is a partial order relation on S, but no two distinct elements of the subset are related by the relation.

anticommutative This is a bad word, the marriage of the Greek preposition ἀντί- (*against, opposed to, opposite to*) and the Latin-based word *commutative*. The equivalent Latin preposition *contra* should have been used at the time of birth. Such absurd words, half Latin and half Greek, are exclusively concocted by people ignorant of both those languages. Dr. Johnson condemns the use of such hybrids as typical of the confused speech of barbarians, conveying by a mixture of signs and grunts ideas that they are unable to get across singly in any one of those ways. Words of this type may eventually gain general acceptance with the multitude; consider, for example, the word *television*, which is of this sort. When used of branches of learning, such as *neuroscience* or *audiology*, they are a heads-up that some quackery may be involved. They are intended by their creators to describe a property that excludes whatever comes after the *anti-*. Thus, a binary operation $*$ is anticommutative if $a * b = -(b * a)$.

antiderivative This word is the union of the Greek preposition ἀντί-, *against, opposed to, opposite to*, and the Latin-based word *derivative*. A function F whose derivative is a second function f is called an *antiderivative* of f. See the comment above on *anticommutativity*.

antilogarithm A number x whose logarithm is y is called the *antilogarithm* of y. The word is compounded of the preposition ἀντί-, *against, opposed to, opposite to,* and the Greek nouns λόγος, *word, reason,* and ἀριθμός, *number*. The use of *anti-* in this word cannot be faulted because *logarithm* is itself a word of Greek origin. The word is a late seventeenth-century offspring of the word *logarithm*, invented by John Napier in 1614.

antinomy The Greek noun ἀντινομία means *a conflict of laws, an ambiguity in the law*. It is derived from the verb ἀντινομέω, *to disobey*, composed of ἀντί-, *against*, and νόμος, *law*. An antinomy is an apparent contradiction, a paradox, such as Russell's antinomy in set theory, which considers the set of all sets that are not members of themselves.

antipodal This is a Greek word for points at the opposite ends of a solid, for example, points at the opposite ends of a sphere. It is composed of the Greek preposition ἀντί- (*against, opposed to, opposite to*), the stem of the Greek noun πούς, ποδός (*foot*), and the stem of the Latin adjectival suffix *-alis*. It is thus macaronic. A pure equivalent would have been *antipodic*.

antisymmetric This is a modern word formed from the Greek preposition ἀντί- (*against, opposed to, opposite to*) and the adjective συμμετρικός, which means *symmetric*.

aperiodic This good word means *not periodic;* it is compounded of *alpha privativum* and the adjective *periodic* (*q.v.*) It would have been *aperiodal* if it had been invented by the same person who invented *antipodal.*

apex This is a Latin word that means the top point of anything, especially that of a pyramid or of a cone.

aphelion This is the point in the earth's orbit when it is farthest from the sun. This point is reached on July 4. It is one of the two points in the orbit where the position and velocity vectors are perpendicular. If r_0 and v_0 are the distance and speed at aphelion, then the constant rate at which the radius vector from the sun to the earth sweeps out area is $r_0 v_0 / 2$. The word is composed of the Greek preposition ἀπό, *from,* and the noun ἥλιος, *the sun, which sees all and hears all.* It was concocted by Kepler on the analogy of *apogee.*

apogee This word is composed of the Greek preposition ἀπό, *from,* and the noun γῆ, *earth.* The Greek adjective ἀπόγαιος (also ἀπόγειος) means *far from the earth.* The *apogee* is the point in the moon's orbit where it is farthest from the earth. The double *e* at the end is the sign of the Greek diphthong ει. Thus, the name of the metrical unit that the Greeks called ὁ σπονδεῖος πούς in their language is in English a *spondee.*

apothem This is the perpendicular distance from the center of a regular polygon to a side. *Apothem,* without the *g,* is the correct spelling. It is the Greek ἀπόθεμα from the verb ἀποτίθημι, *to place* (τίθημι) *from* (ἀπό). The spelling *apothegm* is due to confusion with the different word *apophthegm* from the verb ἀποφθέγγεσθαι, *to speak out;* an *apophthegm* is a pithy saying.

applied This is the Latin past participle *applicatus,* the fourth principal part of the verb *applico, applicare, applicavi, applicatus,* which means *to place to or near, to attach or connect.* It is compounded of the preposition *ad* (*to*) and the verb *plico* (*to fold*). Among the Greeks, to apply one figure to another was to bring the former into contact with the other. applied mathematics is that portion of the subject that serves as the handmaiden, rather than the queen, of science.

approximation The Latin preposition *ad* means *to*, and the superlative adjective *proximus* means *nearest*. From their combination (after the prefix *ad-* was changed to *ap-* by assimilation for the sake of euphony) arose the verb *approximo, approximare, approximavi, approximatus* with the meaning *to be near or draw near to*. To *approximate*, therefore, is properly an intransitive, not a transitive, verb. From its fourth principal part is formed the noun *approximatio, -onis* meaning *a drawing near to*. Thus, an approximation is something near something else. It is a medieval Latin word.

approximoscope This absurdity is the macaronic combination of the Latin *approximo* (see the previous entry) and the Greek σκοπός, *a look-out*, from σκοπέω, *to look, to look out*. It appears, together with some other silly words, in the appropriately titled article "A Farey Tail" in the June/July 2012 issue of the *Notices of the AMS*, vol. 59, no. 6, pages 746–757. Other terms appearing in the article are *garden of visibles*, *Farey comb*, *Farey eye*, and *lightning*. The use of this kind of nomenclature detracts from the majesty of the subject and throws the mantle of absurdity over the scientific content that the essay has. Levity is unbecoming to mathematics. *Procul O, procul este, profani. Farey* is one of those few names that cannot be compounded with certain other nouns without ludicrous results. That such a system of denominating mathematical objects can be adopted is nothing to be marveled at in an age when people give their own children bizarre names.

   The principles that infallibly guide the author to correct practice in the coining of words were described by Horace in a passage (*Epistolarum Liber* II, II [*Ars Poetica*], 110–125) that was taken by Dr. Johnson to be the motto of his *Dictionary of the English Language*. (The brackets enclose lines in the poem but not in the motto.)

> [Ridentur mala qui component carmina; verum
> Gaudent scribentes et se venerantur et ultro,
> Si taceas, laudant quicquid scripsere beati.
> At qui legitimum cupiet fecisse poema,]
> Cum tabulis animum censoris sumet honesti:

Audebit quaecunque parum splendoris habebunt,
Et sine pondere erunt, et honore indigna ferentur.
Verba movere loco; quamvis invita recedant,
Et versentur adhuc intra penetralia Vestae:
Obscurata diu populo bonus eruet, atque
Proferet in lucem speciosa vocabula rerum,
Quae priscis memorata Catonibus atque Cethegis,
Nunc situs informis premit et deserta vetustas.
[Adsciscet nova, quae genitor produxerit usus.
Vehemens et liquidus puroque simillimus amni
Fundet opes Latiumque beabit divite lingua;
Luxuriantia compescet, nimis aspera sano
Levabit cultu, virtute carentis tollet,
Ludentes speciem dabit et torquebitur, ut qui
Nunc Satyrum, nunc agrestem Cyclopa movetur.

[Bad poets are our jest: yet they delight,
Just like their betters, in whate'er they write,
Hug their fool's paradise, and if you're slack
To give them praise, themselves supply the lack.]
But he who meditates a work of art,
Oft as he writes, will act the censor's part:
Is there a word wants nobleness and grace,
Devoid of weight, unworthy of high place?
He bids it go, though stiffly it decline,
And cling and cling, like suppliant to a shrine:
Choice terms, long hidden from the general view,
He brings to day and dignifies anew,
Which, once on Cato's and Cethegus' lips,
Now pale their light and suffer dim eclipse;
New phrases, in the world of books unknown,
So use but father them, he makes his own:
Fluent and limpid, like a crystal stream,
He makes Rome's soil with genial produce teem:
He checks redundance, harshnesses improves
By wise refinement, idle weeds removes;
Like an accomplished dancer, he will seem
By turns a Satyr and a Polypheme;
Yet all the while 'twill be a game of skill,
Where sport means toil, and muscle bends to will.
(Translation by John Conington, M.A., 1825–1869, Corpus
Christi Professor of Latin, Oxford University)

Unqualified people, however, follow not the principles of Horace, but the license allowed by the carelessness produced by modern education. Words formed from word fragments are an especially blameworthy violation of the laws of literacy. In our own profession there has arisen the word *mathlete* formed in this manner from the word *athlete*. The word *athlete* is Greek and means *someone who competes in a contest*. Adding the letter *m* at the beginning produces the absurdity *mathlete*, a student who wins a mathematical competition. *Mathlete* is a low word similar to the silly *guestimate* and *labradoodle*. Another monstrosity of the same kind is *Webinar*. *Twitterverse* is yet another.

Another violation of literary decency is macaronic composition and the coining of macaronic words. Dr. Johnson had the following to say on macaronic compositions. The excerpt concerns Pope's Epitaph for James Craggs, Esq., in Westminster Abbey:

> It may be proper here to remark the absurdity of joining in the same inscription Latin and English, or verse and prose. If either language be preferable to the other, let that alone be used; for no reason can be given why part of the information should be given in one tongue and part in another on a tomb, more than in any other place or any other occasion; and to tell all that can be conveniently told in verse, and then to call in the help of prose, has always the appearance of a very artless expedient, or of an attempt unaccomplished. Such an epitaph resembles the conversation of a foreigner, who tells part of his meaning by words, and conveys part by signs. (Samuel Johnson, *The Lives of the Most Eminent English Poets, with Critical Observations on their Works*, 4 vols., London, 1781, vol. 4, pp. 224–225)

Dr. Johnson was merely expressing in prose form the opinion of Horace expressed in the opening nine verses of the *Ars Poetica* (*Epistolarum Liber* II, 3, 1–9):

> Humano capiti cervicem pictor equinam
> Iungere si velit et varias inducere plumas
> Undique collatis membris, ut turpiter atrum
> Desinat in piscem mulier formosa superne;
> Spectatum admissi risum teneatis amici?
> Credite, Pisones, isti tabulae fore librum

Persimilem, cuius, velut agri somnia, vanae
Fingentur species, ut nec pes nec caput uni
Reddatur formae.

Suppose a painter to a human head
Should join a horse's neck, and wildly spread
The various plumage of the feathered kind
O'er limbs of different beasts, absurdly joined;
Or if he gave to view a beauteous maid
Above the waist with every charm arrayed,
Should a foul fish her lower parts infold,
Would you not laugh such pictures to behold?
Such is the book, that like a sick man's dreams,
Varies all shapes, and mixes all extremes.
(Translation by Philip Francis, 1708–1773)

a priori, a posteriori These Latin prepositional phrases were first used as technical terms in English by George Berkeley in *A Treatise concerning the Principles of Human Knowledge* (1710): "I think arguments *a posteriori* are unnecessary for confirming what has been...sufficiently demonstrated *a priori*" (§22). As the excerpt indicates, when used with nouns, they should come in the predicative, not attributive, position, as befits adverbial phrases modifying an adjective that is understood, for example, "a conclusion [drawn] *a priori*." Nevertheless, in some phrases, they have become, in contempt of Latin grammar, virtual adjectives in English prose, for example, "Descartes' *a priori* proof of the existence of God." Some maintain that those rules of English grammar traceable to Latin, such as the prohibition against splitting infinitives, are illegitimate and may be set aside with impunity; such an attitude is blameworthy. Rules of grammar in the literary world are like protocol in the social world; protocol keeps in their place people who do not know their place. *A priori* means *from what has gone before*; *a posteriori* means *from what has come after*. Conclusions drawn *a priori* are deduced by the method of deductive reasoning from previously assumed axioms and previously proven propositions; they are obtained by the method of mathematics. Those established *a posteriori* are the ones known as a result of the application of the Baconian, that is, experimental, philosophy; they are known as a result of experience. An example may be adduced from the theory of probability. If the events A_1, A_2,... are a partition of the sample space $\textit{δ}$, and B is an event, then the probabilities $P(A_1)$, $P(A_2)$,... are the probabilities assumed *a priori* or the prior probabilities, which are just given, whereas the conditional probabilities $P(A_1|B)$, $P(A_2|B)$,... are the probabilities obtained *a posteriori*, or the posterior or adjusted probabilities calculated after the occurrence of the event B. For the use of these phrases in philosophy, see the article *"a priori / a posteriori"* in Simon Blackburn's *Oxford Dictionary of Philosophy,* Oxford University Press, Oxford and New York, 1994.

Arabic This is the Latin adjective *arabicus* from the noun *arabs, arabis*, an Arab. In the Arabic language, the word is عرب, which comes from a root meaning *dry* and means a dweller of the arid steppes of northern Arabia. In mathematics it is used of the system of numerals *1, 2, 3, 4,*... derived from the Arabs in the time of the Crusades, numerals that they themselves had received from the civilization of the Indus. The word must be capitalized in English; spellings such as arabic, greek, french, etc. cannot be defended.

arbitrary The Latin adjective *arbitrarius* means *uncertain*. It comes from the noun *arbiter*, which means first *witness* and then *umpire, judge, master*. The *arbitrator* was the fellow appealed to for a judgment; his decision was the *arbitrium*. From *arbitrium* there was formed the adjective *arbitrarius* by means of the addition of the adjectival ending *-arius*. Thus, that which has the properties of a final decision reached by a judge was termed *arbitrary*.

arc The Latin word for *bow* is *arcus, arcūs*. It also has the derived meanings of *vault, arch,* or *anything curved*. The word accordingly was applied by Seneca (*Quaestiones Naturales,* Book I, 10, 1) to a piece of the circular arc of the rainbow and from thence to a piece of any curve.

> Similis varietas in coronis est; sed hoc differunt, quod coronae ubique fiunt, ubicunque sidus est, arcus non nisi contra solem....

A like difference occurs in the case of halos; but they differ in this respect, that while halos are found everywhere, wherever there is a star or planet, a rainbow occurs only near the sun....

arccos This is an abbreviation for *arcus cosinūs*, the arc of the cosine. The expression *arccos x* was intended to be interpreted as *the angle* (actually, *the arc*) *whose cosine is x*. The notation *arc cos x* is unobjectionable. See the entry for **cosine**.

arccot This, as well as *arcctn*, is an abbreviation for *arcus [lineae] cotangentis*, the arc of the cotangent [line]. The expression *arccot x* was intended to be interpreted as *the angle* (actually, *the arc*) *whose cotangent is x*, for in a circle of given radius, the angle is known once the subtended arc is known. The *cotangent line* is defined as follows: If one draws a central angle θ in the first quadrant of a unit circle with center at the origin, with the initial side of the angle θ along the horizontal axis, let O be the origin, B the point where the terminal side of θ intersects the circle, A the point where the perpendicular from B intersects the horizontal axis, C the point where the initial side of θ intersects the circle, D the point where the perpendicular to OC at C intersects the terminal line of θ, E the point on the terminal side of θ where the perpendicular to the initial side of θ is of length one, and F the point where this perpendicular intersects the initial side. Then OF is the cotangent line. The notations *arc cot x* and *arcctn x* are unobjectionable. See the entry **cotangent**.

arccsc This is an abbreviation for *arcus [lineae] cosecantis*, the arc of the cosecant [line]. The expression *arccsc x* was intended to be interpreted as *the angle* (actually, *the arc*) *whose cosecant is x*, for in a circle of given radius, the angle is known once the subtended arc is known. The *cosecant line* is defined as follows: If one draws a central angle θ in the first quadrant of a unit circle with center at the origin, with the initial side of the angle θ along the horizontal axis, let O be the origin, B the point where the terminal side of θ intersects the circle, A the point where the perpendicular from B intersects the horizontal axis, C the point where the initial side of θ intersects the circle, D the point where the perpendicular to OC at C intersects the terminal line of θ, and E the point on the terminal side of θ where the perpendicular to the initial side of θ is of length one. Then OE is the cosecant line. The notation *arc csc x* is unobjectionable. See the entry for **cosecant**.

Archimedean The spelling *Archimedian* is incorrect since the *e* is part of the name of the scientist. The name of the greatest mathematician of antiquity was Ἀρχιμήδης, which means *foremost* (ἀρχι-) *in counsel* (μῆδος). The inseparable prefix ἀρχι- comes from the verb ἄρχω, which means *to begin, to be first, to be the leader*. The *Archimedean spiral* is the polar curve with equation $r = k\theta$, where k is any constant; if a point mass P moves out from the pole along the initial line at constant speed s at the same time that the initial line rotates at constant angular speed ω, then the locus of P is $r = (s/\omega)\theta$. Archimedes pointed out (*On Spirals*) that if this curve is given, then one can trisect any angle. Because this spiral cannot be constructed with unmarked straightedge and compass, it was denominated *mechanical*, a term of opprobrium.

Archimedes The story of Archimedes in Plutarch's *Life of Marcellus* is a masterpiece; it was the first biography of a mathematician (if so it might be called) and established the stereotype of mathematicians as absent-minded and eccentric. Archimedes died in 212 B.C., and Plutarch wrote the *Parallel Lives* about three and a quarter centuries later. The translation is that attributed to Dryden.

> ...The land forces were conducted by Appius: Marcellus, with sixty galleys, each with five rows of oars, furnished with all sorts of arms and missiles, and a huge bridge of planks laid upon eight ships chained together, upon which was carried the engine to cast stones and darts, assaulted the walls, relying on the abundance and magnificence of his preparations, and on his own previous glory; all which, however, were, it would seem, but trifles for Archimedes and his machines.
>
> These machines he had designed and contrived, not as matters of any importance, but as mere amusements in geometry; in compliance with King Hiero's desire and request, some little time before, that he should reduce to practice some part of his admirable speculation in science, and by accommodating the theoretic truth to sensation and ordinary use, bring it more within the appreciation of the people in

general. Eudoxus and Archytas had been the first originators of this far-famed and highly-prized art of mechanics, which they employed as an elegant illustration of geometrical truths, and as means of sustaining experimentally, to the satisfaction of the senses, conclusions too intricate for proof by words and diagrams. As, for example, to solve the problem, so often required in constructing geometrical figures, given the two extremes, to find the two mean lines of a proportion, both these mathematicians had recourse to the aid of instruments, adapting to their purpose certain curves and sections of lines. But what with Plato's indignation at it, and his invective against it as the mere corruption and annihilation of the one good of geometry, which was thus shamefully turning its back upon the unembodied objects of pure intelligence to recur to sensation, and to ask help (not to be obtained without base supervisions and depravation) from matter; so it was that mechanics came to be separated from geometry, and repudiated and neglected by philosophers, took its place as a military art. Archimedes, however, in writing to King Hiero, whose friend and near relation he was, had stated that given the force, any given weight might be moved, and even boasted, we are told, relying on the strength of demonstration, that if there were another earth, by going into it he could remove this. Hiero, being struck with amazement at this, and entreating him to make good this problem by actual experiment, and show some great weight moved by a small engine, he fixed accordingly upon a ship of burden out of the king's arsenal, which could not be drawn out of the dock without great labour and many men; and loading her with many passengers and a full freight, sitting himself the while far off, with no great endeavour, but only holding the head of the pulley in his hand and drawing the cords by degrees, he drew the ship in a straight line, as smoothly and evenly as if she had been in the sea. The king, astonished at this, and convinced of the power of the art, prevailed upon Archimedes to make him engines accommodated to all the purposes, offensive and defensive, of a siege. These the king himself never made use of, because he spent almost all of his life in a profound quiet and the highest affluence. But the apparatus was, in most opportune time, ready at hand for the Syracusans, and with it also the engineer himself.

When, therefore, the Romans assaulted the walls in two places at once, fear and consternation stupefied the Syracusans, believing that nothing was able to resist that violence and those forces. But when Archimedes began to ply his engines, he at once shot against the land forces all sorts of missile weapons and immense masses of stone that came down with incredible noise and violence, against which no man could stand; for they knocked down those upon whom they fell in heaps, breaking all their ranks and files. In the meantime huge poles thrust out from the walls over the ships sunk some by the great weights which they let down from on high upon them; others they lifted up into the air by an iron hand or beak like a crane's beak and, when they had drawn them up by the prow, and set them on end upon the poop, they plunged them to the bottom of the sea; or else the ships, drawn by engines within, and whirled about, were dashed against steep rocks that stood jutting out under the walls, with great destruction of the soldiers that were aboard them. A ship was frequently lifted up to a great height in the air (a dreadful thing to behold), and was rolled to and fro, and kept swinging, until the mariners were all thrown out, when at length it was dashed against the rocks, or let fall. At the engine that Marcellus brought upon the bridge of ships, which was called *Sambuca*, from some resemblance it had to an instrument of music, while it was as yet approaching the wall, there was discharged a piece of rock of ten talents weight, then a second and a third, which, striking upon it with immense force and a noise like thunder, broke all its foundation to pieces, shook out all its fastenings, and completely dislodged it from the bridge. So Marcellus, doubtful what counsel to pursue, drew off his ships to a safer distance, and sounded a retreat to his forces on land. They then took a resolution of coming up under the walls, if it were possible, in the night; thinking that as Archimedes used ropes stretched at length in playing his engines, the soldiers would now be under the shot, and the darts would, for want of sufficient distance to throw them, fly over their heads without effect. But he, it appeared, had long before framed for such occasions engines accommodated to any distance, and shorter weapons; and had made numerous small openings in the walls, through which, with engines at a shorter range, unexpected blows were inflicted on the assailants. Thus, when they who thought to deceive the defenders came close up to the walls, instantly a shower of darts and other missile weapons was again cast upon them. And when stones came tumbling down perpendicularly upon their heads, and, as it were, the whole wall shot arrows out at them, they retired. And now, again, as they were going off, arrows and darts of a longer range inflicted a great slaughter upon them, and their ships were driven one against another; while they themselves were not able to retaliate in any way. For Archimedes had provided and fixed most of his engines immediately under the wall; whence the Romans, seeing that indefinite mischief overwhelmed them from no visible means, began to think they were fighting with the gods.

Yet Marcellus escaped unhurt, and deriding his own artificers and engineers, "What," said he, "must we give up fighting with this geometrical Briareus, who plays pitch-and-toss with our ships, and, with the multitude of darts which he showers at a single moment upon us, really outdoes the hundred-headed giants of mythology?" And, doubtless, the rest of the Syracusans were but the body of Archimedes's designs, one soul moving and governing all; for, laying aside all other arms, with this alone they infested the Romans and protected themselves. In fine, when such terror had seized upon the Romans, that, if they did but see a little rope or a piece of wood from the wall, instantly crying out, that there it was again, Archimedes was about to let fly some engine at them, they turned their backs and fled. Marcellus desisted from conflicts and assaults, putting all his hope in a long siege. Yet Archimedes possessed so high a spirit, so profound a soul, and such treasures of scientific knowledge, that though these inventions had now obtained him the renown of more than human sagacity, he yet would not deign to leave behind him any commentary or writing on such subjects; but repudiating as sordid and ignoble the whole trade of engineering, and every sort of art that lends itself to mere use and profit, he placed his whole affection and ambition in those purer speculations where they can be no reference to the vulgar needs of life; studies, the superiority of which to all others is unquestioned, and in which the only doubt can be whether the beauty and grandeur of the subjects examined, or the precision and cogency of the methods and means of proof, most deserve our admiration. It is not possible to find in all geometry more difficult and intricate questions, or more simple and lucid demonstrations. Some ascribe this to his natural genius; while others think that incredible effort and toil produced these, to all appearances, easy and unlaboured results. No amount of investigation of yours would succeed in attaining the proof, and yet, once seen, you immediately believe you would have discovered it; by so smooth and so rapid a path he leads you to the conclusion required. And thus it ceases to be incredible that (as is commonly told of him) the charm of his familiar and domestic Siren made him forget his food and neglect his person, to that degree that when he was occasionally carried by absolute violence to bathe or have his body anointed, he used to trace geometrical figures in the ashes of the fire, and diagrams in the oil on his body, being in a state of entire preoccupation, and, in the truest sense, divine possession with his love and delight in science. His discoveries were numerous and admirable; but he is said to have requested his friends and relations that, when he was dead, they would place over his tomb a sphere containing a cylinder, inscribing it with the ratio which the containing solid bears to the contained.

Such was Archimedes, who now showed himself, and so far as lay in him the city also, invincible...But nothing afflicted Marcellus so much as the death of Archimedes, who was then, as fate would have it, intent upon working out some problem by a diagram, and having fixed his mind alike and his eyes upon the subject of his speculation, he never noticed the incursion of the Romans, nor that the city was taken. In this transport of study and contemplation, a soldier, unexpectedly coming up to him, commanded him to follow to Marcellus; which he declining to do before he had worked out his problem to a demonstration, the soldier, enraged, drew his sword and ran him through. Others write that a Roman soldier, running upon him with a drawn sword, offered to kill him; and that Archimedes, looking back, earnestly besought him to hold his hand a little while, that he might not leave what he was then at work upon inconclusive and imperfect; but the soldier, nothing moved by this entreaty, instantly killed him. Others again relate that as Archimedes was carrying to Marcellus mathematical instruments, dials, spheres, and angles, by which the magnitude of the sun might be measured to the sight, some soldiers seeing him, and thinking that he carried gold in a vessel, slew him. Certain it is that his death was very afflicting to Marcellus; and that Marcellus ever after regarded him that killed him as a murderer; and that he sought for his kindred and honoured them with signal favours.

The reader will have noticed the mistake in the explanation of the story regarding the theorem on the ratio of the volumes of the cylinder, sphere, and cone. The mistake is that of the translators, not of Plutarch, who got it right, for he wrote

λέγεται τῶν φίλων δεηθῆναι καὶ τῶν συγγενῶν, ὅπως αὐτοῦ μετὰ τὴν τελευτὴν ἐπιστήσουσι τῷ τάφῳ τὸν περιλαμβάνοντα τὴν σφαῖραν ἐντός κύλινδρον... (The text is reproduced from the Teubner edition of Plutarch on p. 200 of

Lester H. Lange, "Did Plutarch Get Archimedes' Wishes Right?" *The College Mathematics Journal*, vol. 26, no. 3, May 1995, pp. 199–204.)

It is reported that he begged his friends and relations that after his death they should have engraved on his tombstone the cylinder circumscribing the sphere...

The rediscovery of the tomb of Archimedes, inscribed with the diagram illustrating the theorem, is the subject of a famous passage in Cicero:

XXIII. With the life of such a man [Dionysius the tyrant], and I can imagine nothing more horrible, wretched, and abominable, I shall not indeed compare the life of Plato or Archytas, men of learning and true sages: I shall call up from the dust on which he drew his figures an obscure, insignificant person belonging to the same city [Syracuse], who lived many years after, Archimedes. When I was quaestor, I tracked out his grave, which was unknown to the Syracusans (as they totally denied its existence), and found it enclosed all round and covered with brambles and thickets; for I remembered certain doggerel lines inscribed, as I had heard, upon his tomb, which stated that a sphere along with a cylinder had been set up on the top of his grave. Accordingly, after taking a good look all round (for there are a great quantity of graves at the Agrigentine Gate), I noticed a small column rising a little above the bushes, on which there was the figure of a sphere and a cylinder. And so I at once said to the Syracusans (I had their leading men with me) that I believed it was the very thing of which I was in search. Slaves were sent in with sickles who cleared the ground of obstacles, and when a passage to the place was opened we approached the pedestal fronting us; the epigram was traceable with about half the lines legible, as the latter portion was worn away. So you see, one of the most famous cities of Greece, once indeed a great school of learning as well, would have been ignorant of the tomb of its one most ingenious citizen, had not a man of Arpinum pointed it out. But to come back to the point where I made this digression. Who in all the world, who enjoys merely some degree of communion with the Muses, that is to say with liberal education and refinement, is there who would not choose to be the mathematician than the tyrant?... (Tusculan Disputations V, xxiii, 64–66, translation found in Loeb's Classical Library)

arcsec This is an abbreviation for *arcus [lineae] secantis*, the arc of the secant [line]. The expression *arcsec x* was intended to be interpreted as *the angle* (actually, *the arc*) *whose secant is x*, for in a circle of given radius, the angle is known once the subtended arc is known. The *secant line* is defined as follows: If one draws a central angle θ in the first quadrant of a unit circle with center at the origin, with the initial side of the angle θ along the horizontal axis, let O be the origin, B the point where the terminal side of θ intersects the circle, A the point where the perpendicular from B intersects the horizontal axis, C the point where the initial side of θ intersects the circle, and D the point where the perpendicular to OC at C intersects the terminal line of θ. Then OD is the secant line. The notation *arc sec x* is unobjectionable. See the entry **secant.**

arcsin This is an abbreviation for *arcus sinūs*, the arc of the sine. The expression *arcsin x* was intended to be interpreted as *the angle* (actually, *the arc*) *whose sine is x*. The notation *arc sin x* is unobjectionable and was, in fact, used by Konrad Knopp. See the entry **sine.**

arctan This is an abbreviation for *arcus [lineae] tangentis*, the arc of the tangent [line]. The expression *arctan x* was intended to be interpreted as *the angle* (actually, *the arc*) *whose tangent is x*, for in a circle of given radius, the angle is known once the subtended arc is known. The *tangent line* is defined as follows: If one draws a central angle θ in the first quadrant of a unit circle with center at the origin, with the initial side of the angle θ along the horizontal axis, let O be the origin, B the point where the terminal side of θ intersects the circle, A the point where the perpendicular from B intersects the horizontal axis, C the point where the initial side of θ intersects the circle, and D the point where the perpendicular to OC at C intersects the terminal line of θ. Then DC is the tangent line. The notation *arc tan x* is unobjectionable and was, in fact, used by Konrad Knopp. Continuous random variables are said to observe the *arctangent law* if their probability density function is $f(x) = 2/\pi(x^2 + 4)$. See the entry **tangent.**

area This is the Latin word for a vacant level or open space in a town. Its use as the term of measure in two dimensions is found in Billingsley's *Euclid* (1570).

argsinh, argcosh, argtanh, argcoth, argsech, argcsch These are abbreviations for the Latin *argumentum sinūs hyperbolici, argumentum cosinūs hyperbolici, argumentum tangentis hyperbolici, argumentum cotangentis hyperbolici, argumentum secantis hyperbolici,* and *argumentum cosecantis hyperbolici;* the argument of the hyperbolic sine, cosine, tangent, cotangent, secant, or cosecant. See the entry **argument**. The *argument of the hyperbolic function* is that by means of which the hyperbolic function may be determined. The notation *arcsinh, arccosh, arctanh, arccoth, arcsech,* and *arccsch* is wrong, despite its adoption by the oracle Wikipedia since the inverse hyperbolic cosine has nothing to do with any arc. To write $Sinh^{-1}$ or $sinh^{-1}$, $Cosh^{-1}$ or $cosh^{-1}$, etc. is also bad since it may be justifiably interpreted as referring to the hyperbolic cosecant, the hyperbolic secant, etc. The words should be read *inverse hyperbolic sine, inverse hyperbolic cosine,* etc., not pronounced as written.

argument *Argumentum* is a Latin word for *proof* derived from the verb *arguo, arguer, argui, argutus,* which means *to put in clear light,* from the Greek adjective ἀργός, *bright.* An *argumentum* is that by means of which something else is made clear. The definition of the mathematical term *argument* given by the *Oxford English Dictionary* is excellent: "The angle, arc, or other mathematical quantity, from which another required quantity may be deduced, or on which its calculation depends." If f is a function and $y = f(x)$, then x is the argument of f, from which y may be deduced. The argument of a complex number $a + bi$ in the plane is the angle that the radius vector from the origin to (a,b) makes from the horizontal.

-aris This is the form of the Latin adjectival suffix *-alis* that was appended to noun stems that ended in l to avoid cacophony. English adjectives ending in *-ar* and of Latin origin are usually to be explained as having had this element in their history. See the entry **-alis**.

arithmetic This word comes from the Greek ἀριθμός, *number*, whence the adjective ἀριθμητικός, *having to do with number*, was derived. The Greeks used the phrase ἡ ἀριθμητικὴ τέχνη, *the arithmetic art*, to mean what we would call *number theory*.

ascending The Latin verb *ascendo, ascendere, ascendi, ascensus* means *to go up*. The stem became the English verb *ascend*.

associate *Adsocio, adsociare, adsociavi, adsociatus* is a late Latin verb meaning *to take as a colleague for oneself*, formed by uniting the prefix *ad-* to the verb *socio, sociare*, which means *to unite*. The *d* became an *s* through assimilation.

associative This verbal adjective is formed by adding the adjectival ending *-ivus* to the perfect passive participle *adsociatus* of the verb *adsocio*.

assumption *Adsumo, adsumere, adsumpsi, adsumptus* is a Latin verb meaning *to take to oneself*. Cicero used the word to mean *to state the hypothesis of a conditional sentence*, which hypothesis was called the *adsumptio, adsumptionis*. The ending *-n* in English indicates that this sort of word was taken over from the Latin oblique cases. The *d* became an *s* through assimilation.

asteroid This is meant to mean *star-like*. The Greek word for *star* is ἀστήρ, ἀστέρος. The suffix *-oid* comes from εἶδος, which means *shape*. The adjective ἀστεροειδής, *starry*, became *asteroid* in English. There are two words *asteroid* and *astroid* that have different English meanings but are derived from the same Greek word. An *asteroid* is a piece of junk orbiting the sun between the trajectories of Mars and Jupiter and too small to be called a planet. The *astroid* is the hypocycloid of four cusps. If the base circle has radius *a*, the length of the astroid is $6a$; the area of the region it encloses is $3a^2\pi/8$.

astronomical The Greek word ἀστρονομία meant the science that places the stars (ἀστέραι) into categories (νόμοι). The corresponding adjective was ἀστρονομικός, which came into Latin as *astronomicus*. Someone then erred and added the Latin adjectival ending *-alis* on top of the Greek adjectival ending -ικός, to produce *astronomical*, which word, originally comical, is now sanctioned by immemorial custom. Scott Pelley on the *CBS Evening News* of May 17, 2012, said, "The chance of that happening [*sc.* an asteroid hitting the earth] is probably astronomical," but he no doubt should have said *infinitesimal*.

asymmetric This adjective means *not symmetric* and is formed from *alpha privativum* and the Greek adjective συμμετρικός, which means *symmetric*. A binary relation R on a set X is *asymmetric* if $(x,y) \in R \Rightarrow (y,x) \notin R$ for all $x, y \in X$.

asymptote This is a Greek adjective ἀσύμπτωτος meaning *not falling together with, not intersecting with*. It is formed from the conjunction of *alpha privativum*, the preposition σύν, which means *with*, and the verb πίπτω, *to fall*. The mathematical definition of an asymptote allows a curve to intersect its asymptote, but the etymology of the word rules out such a possibility.

atom This is the stem of the Greek adjective ἄτομος, which means *uncut, unmown, indivisible*. It is the combination of *alpha privitivum* and the noun τομή, *a cutting*, from the verb τέμνω, *to cut*. An *atom* of a sample space is an outcome that is an event of positive probability.

augment The Latin verb *augeo, augere, auxi, auctus* means *to make grow, to increase*. From this verb the noun *augmen, augmenis* meaning *increase* was formed. Because of the similarity of this noun to the suffix *-mentum* often added to make nouns from verbs, an extra, superfluous *t* was added at some later point by a confused person who wanted to make the noun English.

auto This is the neuter singular of the adjective αὐτός, αὐτή, αὐτό, which means *the same, the very*. As a prefix, *auto-* should be used with words of Greek origin; to words of Latin origin one ought to add the prefix *idem-*. The corresponding English prefix is *self-*.

autocorrelation This is a macaronic combination like *automobile*; the first part is the Greek prefix αὐτο-, *self, same*, whereas the second part is the stem of a pretended Latin noun *correlatio, -onis*. The verb *refero, referre, retuli, relatus* means *to bring (fero) back (re-)*, and from its fourth principal part is derived the noun *relatio, -onis* with the meaning *a bringing back, a report*. Neither the ancients nor the medieval scholars made the combination *cum + relatio*, which can be traced back to the sixteenth century in English. The *autocorrelation function* is discussed by Nahin on page 221 of *Dr. Euler's Fabulous Formula*, Princeton University Press, Princeton and Oxford, 2006. The author explains how the autocorrelation of a real valued function is a measure of the similarity of that function with a shifted version of itself.

automorphic This adjective was formed by sociologists on the analogy of *anthropormorphic*. The *Oxford English Dictionary* cites a use of the word by Herbert Spencer in 1873. The same authorities coined the noun *automorphism* to mean the ascribing of one's own characteristics to someone else. The Greek adjective αὐτόμορφος means *self-formed, natural*; it is the combination of the prefix αὐτο- from αὐτός, *self*, and μορφή, *form, shape*. The addition of the adjectival suffix *-ic* was unnecessary, as αὐτόμορφος is already an adjective. There is no Greek word αὐτομορφικός.

automorphism This is a modern word concocted from the Greek αὐτός, *self*, and the noun μορφή, *form, shape*. There is no Greek word μορφισμός, let alone αὐτομορφισμός. An automorphism of a group *G* is an isomorphism of *G* onto itself. See the preceding entry.

auxiliary The Latin noun *auxilium* means *help* in English. From this noun comes the adjective *auxiliaris* meaning *helpful*, whence the English *auxiliary*.

average The correct, modern, definition of this word is that it is the same as the mean or expected value of a random variable. There are those who claim that this word is ambiguous and may correctly be applied to what is properly known as the median, or even to any measure of central tendency whatsoever, but this attitude, so conducive to confusion, can never be sufficiently condemned. The first mathematical use of the word was in 1735 by Bishop Berkeley, and the mathematical meaning is noted in *Johnson's Dictionary* (1755). Schwartzman (*The Words of Mathematics: An Etymological Dictionary of Mathematical Terms Used in English*, The Mathematical Association of America, 1994, p. 31b) suggests a connection with the Arabic عور ('-w-r), which in the fourth conjugation means *to borrow*, and in the tenth conjugation *to lend*, but this is precarious. Both Weekley and the *Oxford English Dictionary* declare the etymology of this word to be uncertain.

axiom The root is the Greek verb ἀξιόω, *to think fit, to require*, whence the noun ἀξίωμα, *a statement thought worthy of acceptance, a requirement*. A mathematical assertion was considered worthy of acceptance, originally, because it was believed to correspond with physical reality, that is, with absolute truth, but nowadays axioms are accepted solely for the purpose of seeing what their logical consequences are. Euclid did not use the word *axiom*, instead calling his geometrical assumptions αἰτήματα, *demands, postulates*, and the nongeometrical ones κοιναὶ ἔννοιαι, *common notions*. It was Proclus (fifth century A.D.), in his commentary on the *Elements*, who first called the latter statements *axioms*, and his decision has prevailed to this day. The medieval Latin translators rendered αἰτήματα by *petitiones, requests*, and κοιναὶ ἔννοιαι by *scientia universaliter communis, knowledge common to everyone*.

Bertrand Russell, in his 1959 interview on the BBC program *Face to Face*, described to John Freeman his first encounter with axioms:

> I was never fond of the Classics. I mean, Mathematics was what I liked. My first lesson in Mathematics I had from my brother, who started me on Euclid. I thought it was the most lovely stuff I had ever seen in my life. I didn't know there was anything so nice in the world.
>
> Can you remember and tell us anything about that first lesson?
>
> Oh, yes. I remember it very well, but I remember it was a disappointment because he said, "Now we shall start with

axioms," and I said, "What are they?" He said, "Oh, they are things you've got to admit although we can't prove them." So I said, "Why should I admit things you can't prove?" And he said, "Well, if you don't, we can't go on." And I wanted to see how it went, so I admitted them *pro tem*.

axis From the Greek verb ἄγω, *to lead*, came the noun ὁ ἄξων, τοῦ ἄξονος, which means *axle*. This came into Latin as *axis* or *assis*. There was never a Greek word ἄξις.

azimuth This is the corruption of the Arabic السموت, pronounced *as-samūt*, the plural of السمت, which means *the way*. It is the way from the zenith to the horizon.

B

barycentric [coordinates] *Barycentric* is an English word invented by combining the Greek adjective βαρύς, *heavy*, with the Greek noun κέντρον, *center*, and then superimposing the stem of the adjectival suffix *-ic* from -ικός.

base, basis This is the Greek word βάσις, which means *foundation*, *something to stand on*. The Arabs translated this into their language as القاعدة, *al-qaida*, which is now the name of a terrorist organization. The base of a number system was originally the number of symbols used in writing integers. The most common base was 10 because we have ten fingers. An *open base* for a topological space (X, $\mathcal{J}$) is a family $\mathcal{J}'$ of open sets in $\mathcal{J}$ with the property that every open set in $\mathcal{J}$ is the union of sets in $\mathcal{J}'$.

Bernoulli numbers The alternative name *Bernoullian numbers* is also fine. The Bernoulli numbers B_i, $i \geq 0$, are defined as the solutions of an infinite system of linear equations formed from the Pascal triangle by ignoring the final 1 in each row:

$$B_0 = 1$$
$$1 + 2B_1 = 0$$
$$1 + 3B_1 + 3B_2 = 0$$
$$1 + 4B_1 + 6B_2 + 4B_3 = 0$$
$$\cdot$$
$$\cdot$$
$$\cdot$$

They were introduced by Jakob Bernoulli in his book *Ars Conjectandi* in 1713, pages 95–98. The passage in question is translated into English in the source books of David Eugene Smith and Dirk Struik. The mathematician Ernst Snapper (1913–2011) once asked me why it was that all books on probability refer to *Jakob Bernoulli*, while all books on number theory refer to *Jacques Bernoulli*. I could not give him the reason. The rest of this note is summarized from some of Snapper's lectures on the subject that I once had the pleasure of attending.

The first few Bernoulli numbers may be calculated as follows:

$$B_0 = 1$$
$$B_1 = -1/2$$
$$B_2 = 1/6$$
$$B_3 = 0$$
$$B_4 = -1/30$$
$$B_5 = 0$$
$$B_6 = 1/42$$
$$B_7 = 0$$
$$B_8 = -1/30$$
$$B_9 = 0$$
$$B_{10} = 5/66$$
$$B_{11} = 0$$

The generating function of the Bernoulli numbers is

$$B_0 + B_1 t/1! + B_2 t^2/2! + B_3 t^3/3! + B_4 t^4/4! + \ldots$$

The series has radius of convergence 2π and equals $t/(e^t - 1)$ when convergent.

    The Bernoulli numbers have proved useful in solving various problems in number theory, the first of which is to find a formula for $S_m(n)$, the sum of the mth powers of the first $n - 1$ positive integers. Consider the following series:

$$e^{0t} = 1$$

$$e^t = 1 + t + t^2/2! + t^3/3! + \ldots$$

$$e^{2t} = 1 + 2t + 2^2 t^2/2! + 2^3 t^3/3! + \ldots$$

$$\cdot$$
$$\cdot$$
$$\cdot$$

$$e^{(n-1)t} = 1 + (n-1)t + (n-1)^2 t^2/2! + (n-1)^3 t^3/3! + \ldots$$

Upon addition, we get

$$1 + e^t + e^{2t} + \ldots + e^{(n-1)t} = n + S_1(n)t + S_2(n)t^2/2! + S_3(n)t^3/3! + \ldots$$

The formula for the partial sum of a geometric series and the fact that $n = 1 + (n-1) = 1 + S_0(n)$ give us

$$1 + e^t + e^{2t} + \ldots + e^{(n-1)t} =$$

$$(e^{nt} - 1)/(e^t - 1) = 1 + S_0(n) + S_1(n)t + S_2(n)t^2/2! + S_3(n)t^3/3! + \ldots$$

So

$$1 + S_0(n) + S_1(n)t + S_2(n)t^2/2! + S_3(n)t^3/3! + \ldots =$$

$$(e^{nt} - 1)/(e^t - 1) = [t/(e^t - 1)][(e^{nt} - 1)/t] =$$

$$[B_0 + B_1 t + B_2 t^2/2! + B_3 t^3/3! + \ldots][n + n^2 t/2! + n^3 t^2/3! + n^4 t^3/4! + \ldots].$$

If we write $_nC_k$ for "n choose k," then, by comparison of series, it follows that

$$S_m(n) =$$

$$[n^{m+1} + {}_{m+1}C_1 B_1 n^m + {}_{m+1}C_2 B_2 n^{m-1} + \ldots + {}_{m+1}C_m B_m n]/(m+1).$$

Thus, to calculate $1^{10} + 2^{10} + 3^{10} + \ldots + 1000^{10}$, one needs only to know the first ten Bernoulli numbers. Bernoulli said that it took him less than seven and a half minutes to calculate that the sum in question was equal to 91,409,924,241,424,243,424,241,924,242,500; this is the only instance where the reader might ever need to use the word *nonillion*.

beta This is the second letter of the Greek alphabet, used in mathematics as a name for a function of Euler and a probability distribution. The beta function is defined by

$$B(p,q) = \Gamma(p)\,\Gamma(q)/\,\Gamma(p+q),$$

where p and q are positive real numbers and Γ is the gamma function. The origin of the function lies in the study of the integral of $f(x) = x^{p-1}(1-x)^{q-1}$ over the interval $[0,1]$. The value is $B(p,q)$.

bi- The prefix *bi-* is a Latin abbreviation of the adverb *bis*, which means *twice*.

bicompact This word is the compound of the prefix *bi-* and the adjective *compact*. *Compactus* is the fourth principal part of the verb *compingo, compingere, compegi, compactus, to put together, to construct*, itself formed from the preposition *cum, with*, and the verb *pango, pangere, panxi, pactus, to fasten, fix, drive in, compose, write*.

biconditional This adjective is compounded of the prefix *bi-* and the adjective *conditional*. From the Latin verb *condico, condicere, condixi,* *condictus, to make an arrangement with*, comes the noun *condicio, -onis, an arrangement, agreement, condition, stipulation*. To its stem was added the adjectival suffix *-alis* to produce *conditionalis*, which has the meaning *pertaining to a condition*.

bicontinuous This adjective is compounded of the prefix *bi-* and the adjective *continuous*. The Latin adjective *continuus, connected, hanging together, unbroken*, comes from the verb *contineo, continere, continui, contentus, to hold (teneo) or keep together (cum)*.

bicorn This noun is compounded of the prefix *bi-* and the stem of the Latin noun *cornu*, which means *horn*. Hence Sylvester, who made this word on the analogy of *unicorn* in 1864, intended it to mean *an animal with two horns*. It is the name of the plane curve defined by the equation $(x^2 + 2ay - a^2)^2 = y^2(a^2 - x^2)$, where $|x| \leq |a|$.

bifolium This noun is compounded of the prefix *bi-* and the Latin word *folium*, which means *leaf*. Hence, the *bifolium* is a curve with two leaves. The word is a creation of Kepler, the name for the curve in the polar plane with equation $r = a \sin \theta \cos^2 \theta$. The area of the enclosed region is $a^2 \pi / 32$. The *bifolium* is the pedal curve of the *deltoid* with respect to a vertex.

bifurcate The Latin adjective *bifurcus* is composed of the prefix *bi-* (the Latin abbreviation of the adverb *bis, twice*) and the noun *furca, fork*, from *fero, to carry*. Hence *bifurcus* means *two-pronged*. This word, like *parameter*, is now being used in a bizarre manner by unmathematical authorities. On the May 2, 2012, episode of the *O'Reilly Factor* on the Fox TV station, talking head Dick Morris said, "We have to bifurcate between anti-terrorism and nation building."

biharmonic The prefix *bi-* is a Latin abbreviation of the adverb *bis, twice*. It should therefore be prefixed only to Latin words or words of Latin origin. *Harmonic*, however, is of Greek origin, the stem of the adjective ἁρμονικός with the meaning *pertaining to harmony*. An expert would have made the word *disharmonic* since δίς is the Greek equivalent of the Latin *bis*. The problem with δίς, however, is that its transliteration *dis* is the Latin inseparable suffix that indicates separation or negation, and it is this latter meaning that will occur to the mind of the general public.

biholomorphic The article "What is a Biholomorphic Mapping?" appeared in the June/July 2012 issue of the *Notices of the AMS*, vol. 59, no. 6, pages 812–814. The first paragraph defines the term:

> Let $C^n = C \times \cdots \times C$ denote a complex Euclidean space, and let $D_1, D_2 \subset C^n$ be domains. A mapping
>
> $$f(z_1,\ldots,z_N) = (f_1,\ldots,f_2){:}D_1 \to D_2$$
>
> is *holomorphic* if each of the coordinate functions f_j is holomorphic. If f is one-to-one and onto, then there is an inverse function $f^{-1}{:}\ D_2 \to D_1$, and this may be shown to be holomorphic. In this case, we say that f is *biholomorphic*.

It is clear that the *bi-* here is meant to indicate *in both directions*, as in the case of *bijection*.

bijection The prefix *bi-* is a Latin abbreviation of the adverb *bis, twice*. The Latin verb *iacio, iacere, ieci, iactus* means *to throw*. In compounds, the *a* of the fourth principal part changes to *i*; for example, consider *reicio, reicere, reieci, reiectus*, which means *to reject*. Neither the concoction *jection* nor the Latin *iectus* has any meaning by itself. Nor is there a Latin verb *biicio*, for the Romans did not attach the prefix *bi-* to verbs to make new words. The word *bijection* is therefore low, very low. *Injection* and *surjection*, however, are formed on good analogy since the prepositions *in* and *super* were commonly used as prefixes, so that there were good words *inicio, inicere, inieci, iniectus* and *supericio, supericiere, superieci, superiectus*, from whose fourth principal parts the nouns *injection* and *surjection* were formed, the latter after the mediation of French turned *super* into *sur*.

bilinear This word is the compound of the prefix *bi-*, the Latin abbreviation of the adverb *bis, twice*, and the adjective *linear*, from the Latin adjective *linearis, linear*.

billion This word is derived from the Latin prefix *bi-* (the abbreviation for *bis, twice*) and the word *mille, a thousand*. See **million** below.

bimodal This word is compounded of the Latin *bi-*, from the adverb *bis, twice*, and the medieval Latin adjective *modalis*, formed by adding the adjectival suffix *-alis* to the stem of the noun *modus, a measure or standard of measurement*.

binary The Latin adjective *bini, binae, bina* means *twofold*. From this root there was formed the adjective *binarius*, referring to something that contains twos or consists of twos, whence emerged the English adjective *binary*.

binomial The prefix *bi-* is a Latin abbreviation of the adverb *bis, twice*. It should therefore be prefixed only to Latin words or words of Latin origin. The Greek noun νόμος means *rule* or *law*. Some claim that the word is legitimate because the second component is from the Latin *nomen, name*, with the adjectival ending *-alis* added, but this is unlikely, for then how does one explain the absence of the second *n*? What happened was that the Latin adjectival ending *-alis* was illiterately appended to the Greek noun, and the word became legitimate through its adoption by Newton, from whose authority there is no appeal.

binormal The *norma* was a Roman tool used to check that an angle was right. To this the adjectival ending *-alis* was appended to produce the adjective *normalis, right-angled*. One then added the prefix *bi-* from *bis, twice*, and, lo, one has the word *binormalis* whence the English *binormal*. The Latin form *binormalis* is a modern creation and never actually existed except as a technical term of modern mathematics.

bipartite The classical Latin adjective *bipartitus, -a, -um* means *divided in two* and is derived from the prefix *bi-* (from *bis, twice*) and the fourth principal part of the verb *partio, partire, partivi, partitus*, which means *to share out, distribute, divide*. A *bipartite graph* is defined by

Finkbeiner and Lindstrom (*A Primer of Discrete Mathematics*, W. H. Freeman and Company, New York, 1987, p. 219): A bipartite graph G is a non-empty graph whose vertex set V is partitioned into two non-empty sets L and R, where $V = L \cup R$ and $L \cap R = \emptyset$, such that each edge of G connects a vertex of L and a vertex of R.

biquadratic A *numerus biquadratus* was a number of the form x^4 for some number x. It was so named because it was produced by squaring twice. The later appending of the Greek suffix *-ic* was a mistake; *biquadrate number* would have been better.

bisect This word is composed of the Latin prefix *bi-* (from *bis, twice*) and the syllable *sect* from the verb *seco, secare, secui, sectus, to cut*, whence came the iterative *secto, sectare, to keep on cutting*. The word is an invention of the seventeenth century. Euclid used the phrase δίχα τεμεῖν, *to cut in halves*. Of the Latin translators, Boëthius wrote *in duas aequales dividere partes* (to divide into two equal parts), while Adelard wrote *in duo media dividere* (to divide into two halves).

blogging (mathematical) This is ugly computer lingo. I am tempted to say that it is derived from the Latin verb *bloggo, bloggare*.

bonus-malus This is Latin for *good-bad* and is the name of an insurance system according to which a policy holder is rewarded if he is good (makes no claims) or punished if he is bad (makes claims). The name is ridiculous because it is a pretentious modern invention. Latin phrases abound in the law because *de minimis non curat lex, habeas corpus, quantum meruit*, and the like have their origins in the time when courts conducted their business in Latin, but such phrases should not be made up clumsily where they are out of place.

brachistochrone This is from the Greek βράχιστος χρόνος, *least time*. It is the curve along which a point mass will fall in the least time from one point to another not directly beneath it. It is a mistake to imagine that the brachistochrone connecting two points is always half an arch of a cycloid; this can only be the case if the angle between the line connecting the two points and the vertical is $\pi/2$. The spelling *brachystochrone* is found in the work of Jakob Bernoulli; nowadays it would be called a mistake and marked wrong since the Greek word has an *iota* (ι = *i*), not an *upsilon* (υ = *y*), in the superlative degree; the *upsilon* appears in the positive degree βραχύς. Newton and Leibniz solved the problem posed by Jakob Bernoulli (1654–1705) and proved that the cycloid (*q.v.*) is the brachistochrone; they did so by assuming that the point mass falls as if it were a light ray, which moves according to Fermat's principle, so as to minimize time.

C

c This is the symbol for the cardinal number of the continuum, although the Hebrew ℵ is also used.

calculate The Latin word for *stone* is *calx, calcis*. The addition of the ending *-ulus* to the stem produces the diminutive *calculus*, which means *a small stone* or *pebble*. As a medical term, it is used of bladder, gall, and kidney stones, and even the gritty accumulation on the teeth. Since such pebbles were used as counters in counting, the verb *calculo, calculare, calculavi, calculatus* came into existence with the meaning *to count*. The verb *calculate* is derived from the fourth principal part of this verb, from which is also deduced the noun *calculatio, calculationis* and our noun *calculation*.

calculus This is the Latin word for *a small stone*. See the previous entry. It came to be used by the late seventeenth-century mathematicians as a technical term for any theory that laid the foundations of a general method to calculate the solutions of certain types of problems and then, κατ᾽ ἐξοχήν, to those theories that solved the problems of tangent lines (differential calculus) and quadrature (integral calculus). In modern times *calculus* is the standard introduction to the higher mathematics, and it will remain so until innovators sweep it away with everything else.

calendar mathematics The Greek verb καλέω means *to call*, and from it the Romans derived the name *kalendae, -arum*, the *kalends*, for the first day of their months. The Greeks did not use the *kalends*, so the expression *Greek kalends* came to mean a day that would never come; it is equivalent to the Hebrew expression *when the Messiah comes*.

At the time of Julius Caesar (100–44 B.C.), it was obvious to the world that the days of the year were not occurring in the season originally intended; for example, the first day of spring was falling in June. The error was due to an ancient mistake in estimating the time required for the earth to make one revolution about the sun. The mathematician Sosigenes advised the Perpetual Dictator that the length of the solar year was 365 days and 6 hours. On the basis of this calculation, Caesar introduced in 46 B.C. the calendar that bears his name, and which continues to survive in the Orthodox Church.

The advance of knowledge has since determined the length of the solar year to be 365 days, 5 hours, 48 minutes, 46 seconds. This means that Caesar's year is 11 minutes, 14 seconds too long. The accumulated error amounts to one day every 128.1899 Julian years. As a result, if left uncorrected, Christmas would eventually be celebrated in spring rather than at the beginning of winter. The Supreme Pontiff Gregory XIII (1572–1585) determined to distinguish his pontificate by correcting the calendar of Caesar. There were two problems to be solved. First, he had to correct the accumulated error, which amounted to 12.7 days since 46 B.C. Also, it was necessary to make some change to prevent the error from accumulating again. Since the year 46 B.C. was of no importance for the Catholic religion, the pope determined to restore the situation to where it had been in A.D. 325, the year of the Ecumenical Council of Nicaea. From 325 to 1582 there had elapsed 1,257 years, and the error accumulated during those years amounted to 9.805764 days. It was determined to fix this problem by skipping the 10 days between October 4 and October 15, 1582, a period during which the ecclesiastical calendar had no vital feasts. To prevent the error from reaccumulating, it was determined to skip 3 leap years every 4 centuries, a convenient approximation to the exact error of 3 years every 384.56973 years. It was therefore decreed that centurial years

(years ending in 00) would henceforth be leap years only if divisible by 400. As a result, the Julian calendar fell farther behind the reformed, "Gregorian," calendar in 1700, 1800, and 1900.

The Gregorian calendar is not perfect; the Gregorian year is 26 seconds longer than the actual time that the earth takes for one revolution around the sun. This error amounts to one day every 3,323 years. Thus, another reform of the calendar will be necessary, but may be put off "ad Kalendas graecas."

Consider the date *day d of month m of year 100Y + N*, where $0 \leq Y$ and $0 \leq N \leq 99$. Dates in January and February must be considered to have fallen in the previous year. The formula to find the day of the week for a date in the Julian calendar is

$$J(m) + d - C + N + [N/4] \quad (mod\ 7)$$

where *J(m)* is the function

J(March)	=	*0*
J(April)	=	*3*
J(May)	=	*5*
J(June)	=	*1*
J(July)	=	*3*
J(August)	=	*6*
J(September)	=	*2*
J(October)	=	*4*
J(November)	=	*0*
J(December)	=	*2*
J(January)	=	*5*
J(February)	=	*1*

and the numerical answers are to be interpreted according to the table

Sunday	0
Monday	1
Tuesday	2
Wednesday	3

Thursday	4
Friday	5
Saturday	6

The formula to find the day of the week for a date in the Gregorian calendar is

$$d + G(m) - 2C + N + [N/4] + [C/4] \qquad (mod \ 7)$$

where $G(m)$ is the function

$G(March)$	=	2
$G(April)$	=	5
$G(May)$	=	0
$G(June)$	=	3
$G(July)$	=	5
$G(August)$	=	1
$G(September)$	=	4
$G(October)$	=	6
$G(November)$	=	2
$G(December)$	=	4
$G(January)$	=	0
$G(February)$	=	3

(For the derivation of the formulas, see David M. Burton, *Elementary Number Theory*, sixth edition, McGraw-Hill, pp. 122–127.)

In the Catholic countries of Europe, Thursday, October 4, 1582, was the last day of the Julian calendar. It was followed by Friday, October 15, 1582, the first day of the Gregorian calendar.

In the United Kingdom of Great Britain and Ireland, the last day on which the Julian calendar was used was September 2, 1752. It was followed by September 14, 1752, the day on which the Gregorian calendar was adopted. During the period 1582–1752, the Julian date was denominated *Old Style* (O.S.) and the Gregorian date *New Style* (N.S.). George Washington was born February 11 Old Style, which was February 22 New Style. This was often abbreviated February 11/22. The year of his birth was given 1721/1722, since it was not then the universal custom to call January 1 the beginning of the new year. The Roman calendar considered March the first month, and this practice was reinforced at the advent of Christianity by the occurrence of the Feast of the Annunciation, the observation of the conception of Christ, on March 25.

In Russia, the last day on which the Julian calendar was used was January 31, 1918. The next day was February 14, 1918, the day the Gregorian calendar was adopted.

cancel The noun *cancer, cancri* is the *Crab*, the sign of the Zodiac in which the sun is to be found on June 21 in the northern hemisphere, the day of the summer solstice, when the sun (*sol*) is at its greatest distance from the celestial equator and there appears to be a stopping (*statio*) of its movement; this noun has the plural diminutive *cancelli, -orum*, which means *a lattice, enclosure, grating, grate, balustrade, bars, railing*, and then *the design of the mark × used to obliterate a mistake in a manuscript*. There developed from this the denominative verb *cancello, cancellare, cancellavi, cancellatus* with the meaning *to make lattice-wise, to cross out*, and from its stem is derived the English verb *cancel*. Its past participle is *cancelled*, if we are to follow Dr. Johnson, or *canceled*, if we are to follow Noah Webster. Since the former was more cultured than the latter, the correct spelling is *cancelled*.

canonical The Greek noun κανών, κανόνος originally meant a *rod, bar*, or *carpenter's rule*, whence it developed the meaning *rule, standard*. The corresponding Greek adjective is κανονικός. Someone took this and unnecessarily superimposed the Latin adjectival ending -*alis* to produce *canonicalis*, from which we get the English *canonical*.

cant The Latin verb *cano, canere, cecini, cantus* means *to sing*, and from it is derived the frequentative verb *canto, cantare, cantavi, cantatus*, from whence we have our noun *cant*. Weekley defines it best as *slang, humbug, the whining speech of beggars*. English holds the position once held by Latin as the language in which those books are written that are intended for a universal audience. As a result of this distinction, many English-speaking people do not trouble to study foreign languages and are unable to comment on the etymologies of the words they use. Furthermore, the books that formerly provided the substance of liberal education for an Englishman or American, the works of people who were the most competent and accomplished in their use of words, are no longer part of the curriculum of colleges, where the study of more modern material preoccupies the instructors. This has resulted in a standard of English usage among the faculty of liberal arts colleges that would have astonished a milkmaid or a paper boy of the Enlightenment.

In the following paragraphs, I consider the problem of the low type of English usage that has become commonplace in prose written by mathematicians. The role of a perpetual complainant is as unsuccessful as it is irksome; nevertheless, the public will generally allow a man to say what he likes, provided that he keeps away from the behavior censured by Lord Chesterfield:

> Deep learning is generally tainted with pedantry, or at least unadorned by manners. (Lord Chesterfield, *Letters Written by the Late Right Honourable Philip Dormer Stanhope, Earl of Chesterfield, to His Son, Philip Stanhope, Esq., Late Envoy Extraordinary at the Court of Dresden,* J. Dodsley, London, 1774, vol. I, p. 450)

The freefall of English that took place in the twentieth century resulted in the extinction of good style in the language of mathematics education. In August 1991, I wrote a letter to the editor of the *Notices of the American Mathematical Society* to complain about the style of English used in two recent reports, *viz.,* "Moving Beyond Myths" in the *Notices of the AMS*, July/August, 1991, pages 545–559, and "What Works: Building Natural Science Communities—A Plan for Strengthening Undergraduate Science and Mathematics" (the report of the Project Kaleidoscope Committee). The letter was duly published on page 1085 of the November 1991 issue (vol. 38, no. 9). The following list of blameworthy words and phrases quoted from the reports and offensive to learned ears was included in my letter: statewide mathematics articulation, to replicate effective intervention programs, to enable students to interactively understand, the goal for each experience [*experience* is a new word for *course*], to mainstream students, to remediate students, to sensitize teaching assistants, to educate intending teachers, pipeline population, a lens...polished by their own education, harmful myths about mathematics metastasize to the body politic, interest payments on the deficit of scholarly maturity balloon college enrollments, hands-on curriculum, hands-on learning, hands-on approach, hands-on research, hands-on learning experience, hands-on experiments, hands-on connections, hands-on workshops, hands-on pedagogy, hands-on program, lean and lab-rich, faculty enhancement activities, mathematics…is enhanced, enhance the learning community, set of K–12 experiences, laboratory experiences, hands-on and lab-rich experiences, research experiences, upper-class students will socialize lower-level students, kinesthetic experiences in which students use proprioceptive senses, student-led educational experiences, science experiences, enmeshing the teacher in a laboratory setting, empowering learners, may not be informed by a clear understanding, filtering action, a critical pump in the career pipeline, science and mathematics pipeline, portrait of leakage from the science pipeline, disaggregative enterprise, disaggregated by gender, gender make-up, degrees by gender, facilitators, capstones, clusters, gatekeeper courses, varied menu of courses, upper-class students will socialize lower-level students, spaces [that is, rooms], shape the spaces, etc., etc.

Examples such as these may be multiplied without end to prove the point that the type of English prose found in academia is low, very low, and that mathematicians are as guilty as any other group. It is a major problem of the American educational system.

Upon the publication of my rant in the *Notices*, I received the following letter from Serge Lang:

15 November 1991

Dear Lo Bello,

I just saw your letter to the editor about the garbage language used in pretentious reports being shoved on us by the top wheels in the business. I want to congratulate you. I have had the same reaction. That stuff makes me puke. And I do not agree that these people have something important to say. That remains to be seen.

Serge Lang

That Project Kaleidoscope is alive and well and still up to its old tricks is clear from the prose of its latest mass mailing given below:

Next Generation STEM Learning: Investigate, Innovate, Inspire
November 8–10, 2012
Kansas City, Missouri
Proposals Due March 19, 2012

"Next Generation STEM Learning: Investigate, Innovate, Inspire" will focus on how colleges, community colleges, and universities of all sorts can articulate, expand, measure, and track what works to advance students' achievement of key learning outcomes, emphasizing scientific literacy, quantitative reasoning, analytical thinking, and visionary leadership.

The conference also will feature evidence-based practices that address the urgent need to help underserved students pursue and succeed in STEM courses and programs. It will feature interdisciplinary learning experiences; alignment of program goals for STEM student success across the K–16 continuum; strategies for institutional change that enable twenty-first-century learning; and new approaches in the curriculum and cocurriculum to provide real-world experiences.

A new report from Georgetown University's Center on Education and the Workforce highlights the need for more college graduates, from all majors and disciplines, who have "STEM competencies" for employment and for lifelong well-being. The multi-faceted, unscripted, and borderless challenges of the future—food and water, energy, disease prevention, economic disparities, climate change, conflict and nuclear proliferation—urgently need attention in both society and in higher education. What must educators do to assure that all students who go to college—regardless of major—graduate with the knowledge and agency to meet and to lead productive lives amid these challenges?

AAC&U and Project Kaleidoscope invite proposals that examine and advance the next generation of STEM learning—education that is integrative, links STEM learning with campus and community, develops innovators who can bring new discoveries to market, and inspires collaboration and leadership for a better world.

A more recent report ("Is Moore Better (in Precalculus)?" in *Notices of the AMS*, August 2011, vol. 58, no. 7, pp. 963–965) brings us up to date on the current cant: honors section of precalculus, inquiry-based learning, quasi-experiment, self-efficacy of MMM students, MMM trigger, to assess students' grade self-efficacy, task-specific self-efficacy, detailed grading rubric, gentle discovery method, post-hoc tests.

It is a common occurrence that mistakes, once introduced into a language, become standard over time according to the rule that *the voice of the people is the voice of God*. Indeed, the *norma loquendi* is our only guide. Such changes may be so numerous that a new language is produced in this manner, by the accumulation of mistakes. The best policy is to follow the practice of the best authors, who will decide if a new word or phrase is felicitous or not. This is as natural and reasonable a strategy as to follow the advice of the best doctors in matters of health. That the fund of permissible words is determined by the usage of the best authors was a precept of Lord Chesterfield to be found in one of his letters to his son, whom he was advising in the matter of Latin composition:

> Whenever you write Latin, remember that whatever word or phrase which you make use of, but cannot find in Cesar, Cicero, Livy, Horace, Virgil, and Ovid, is bad, illiberal Latin though it may have been written by a Roman. (Lord Chesterfield, *Letters Written by the Late Right Honourable Philip Dormer Stanhope, Earl of Chesterfield, to His Son, Philip Stanhope, Esq., Late Envoy Extraordinary at the Court of Dresden*, J. Dodsley, London, 1774, vol. I, p. 342, Letter CXXXII)

The advice to read the best authors holds for mathematicians as well as any other class of people.

> It is important to read original writing of great mathematicians. (S. S. Chern, quoted by Jun Li, "Read Classical, Chern Told Us," *Notices of the AMS*, vol. 58, no. 9, p. 1244)

cards The Greek noun χάρτης means a leaf of paper made from the separated layers of papyrus; from the Latin transliteration *charta* there developed the Italian *carta* with the meaning *paper*, the French *carte*, and the English *card*. The calculation of the probabilities of various hands in card games has played an important role in the history of probability, a history that may be studied with profit from the pages of Isaac Todhunter, *A History of the Mathematical Theory of Probability*.

cardinal *Cardo, cardinis* is the Latin word for the *hinge* of a door. There was then appended the suffix *-alis* to produce the adjective *cardinalis*, which was applied to someone who was as important in his profession as a hinge is important to its door. This adjective eventually was restricted to important clergymen in the Roman Catholic Church, but it never quite lost its original meaning of *important*, and this is how it came to be applied to the positive integers, the most important of numbers.

cardioid This word means *heart-shaped*; καρδία is the Greek word for *heart*, and the ending *-oid* is derived from the noun εἶδος, *shape*. The plane curve of this name is a special case of the *limaçon* of Pascal, *q.v.*

Cartesian The Latin name of René Descartes (1596–1650) was Renatus Cartesius. From *Cartesius* one forms the adjective *Cartesianus, -a, -um*, whence we get the English word *Cartesian*. The *Cartesian philosophy* is that approach to life that starts with the principle *de omnibus dubitandum*, that one should doubt everything that cannot not be demonstrated by mathematical argument from clear and distinct principles. The first theorem is *Cogito ergo sum*, I think, therefore I am. The axiom of Descartes, that God does not deceive us, is a *circulus vitiosus*. The most famous portrait of any mathematician is that of Descartes by Frans Hals (1580–1666), on permanent loan to the Royal Museum of Fine Arts in Copenhagen from the Ny Carlsberg Glypotek.

Descartes is the perfect example of a mathematician who understood the meaning of his profession. An examination of *The Discourse of Method* is sufficient for those who are content to know something about Descartes and his philosophy of mathematics rather than everything about them. The book was dedicated to the faculty of the Sorbonne as an insurance policy. In 1937 the French Republic issued a postage stamp to commemorate the three hundredth anniversary of this book, but with the title incorrectly given as *Discours sur la Méthode*. A learned authority brought the mistake to the attention of the Post Office, which withdrew the inaccurate stamp and issued a corrected edition with the title *Discours de la Méthode*. As a result, the first printing, with the mistake, is more expensive to obtain than the corrected version.

The following discussion of the contents of the *Discourse* is taken, with very few minor changes, from my article "Descartes and the Philosophy of Mathematics" published in *The Mathematical Intelligencer*, vol. 13, no. 3, 1991, pages 35–39. The translations are those by Haldane and Ross in the series *Great Books of the Western World*, vol. 31.

The Discourse of Method appeared anonymously, in case any of its doctrines should offend the authorities. It was written in French to underline the fact that it was revolutionary and had something to offer even those people who could not read Latin, for Descartes would not have agreed with Schopenhauer, who wrote in his *Essay on the Study of Latin*:

> If a man knows no Latin, he belongs to the vulgar, even though he is a virtuoso on the electric machine and has the base of hydrofluoric acid in his crucible. (Translation by T. Bailey Saunders)

The title of the work, *The Discourse of Method*, emphasized that Descartes was offering a plan, a well thought out systematic way of acquiring knowledge and then of organizing that knowledge into science. Without the discipline of a method, one could not expect to find the truth. Descartes divided the *Discourse* into six parts so that, he said, his readers could take it leisurely in six installments; however, it is not so long, a mere fifty printed pages. It is a masterpiece of seventeenth-century French prose.

In the first part, Descartes tells how, having studied the usual subjects at school and having travelled over much of Europe to read what he calls "the great book of the world," he had concluded that among "the diverse actions and enterprises of all mankind, I find scarcely any which do not seem to me vain and useless." He therefore decided to turn his mind in on itself and to make himself the object of his study. He was more at home in and by nature more suited to the mental world of ideas rather than the physical world without. He gave evidence of the Platonic predilection for mathematics and noted that of all his school studies,

> Most of all I was delighted with Mathematics because of the certainty of its demonstrations and the evidence of its reasoning.

So, the key to understanding Descartes is that he liked mathematics and that mathematics appeared to him not just one subject among many, not even first among equals, but definitely special.

In Part II, he told of his mystical experience in the stove-heated room, where God appeared to inspire him to begin from scratch:

> As regards all the opinions which up to this time I had embraced, I thought I could not do better than endeavor once and for all to sweep them completely away

and start all over. Descartes was one of those people who are obsessed with wanting to be absolutely certain. Such people must almost surely be disappointed, and Descartes was careful not to recommend his plan for public consumption:

> The simple resolve to strip oneself of all opinions and beliefs formerly received is not to be regarded as an example that each man should follow.

He thought, though, that he might be the exception and end up the better for it, and he was at least sure that in going his own way he would not succumb to those errors that mankind had adopted by unanimous consent:

> The voice of the majority does not afford a proof of any value in truths a little difficult to discover, because such truths are much more likely to have been discovered by one man than by a nation.

Descartes then went on to explain the method of four parts that he had adopted as an infallible procedure for discovering the truth. He came upon it by observing how mathematicians go about their art; mathematics for him provided the correct method of reasoning and seeking for truth in all subjects. The four parts are:

1) To accept nothing as true that he did not clearly recognize to be so;

2) To divide and conquer; to break each big problem up into many smaller ones;

3) To proceed mathematically in solving the smaller problems, that is, from the simplest to the more complex, one at a time according to their order;

4) To check all his work to catch any error of omission or commission.

This method was sure to work, he believed, because

> Those long chains of reasoning, simple and easy as they are, of which geometricians make use in order to arrive at the most difficult demonstrations, had caused me to imagine that all those things which fall under the cognizance of man might very likely be mutually related in the same fashion.

He concluded this section by observing that he was twenty-three years old when he came up with this plan.

Descartes began Part III by observing that because he could not postpone living until he arrived at the truth he was after, he determined to live for the time being according to a reasonable moral code, which also had four parts:

1) To obey the laws and customs of his country, and to adhere to its religion;

2) To be firm and resolute in doing something after having decided to do it;

3) To try always to conquer himself rather than fortune, and to alter his desires rather than try to change the order of the world;

4) To review all the occupations of men in this life in order to determine the best for him, but meanwhile to continue in his own, *viz.*, thinking.

He then described how in his travels he viewed all the comedies that the world displays before withdrawing to Holland to live as quietly as a hermit in deserts the most remote.

In Part IV, Descartes explained that although he could doubt everything else, he could not doubt that he who was thinking existed, and he arrived at the first result of his philosophy, COGITO ERGO SUM—*I think, therefore I am*, the most famous sentence in philosophy. He then proceeded to the highest speculations:

> I saw from the very fact that I thought of doubting the truth of other things, that it very evidently and certainly followed that I was; on the other hand if I had only ceased from thinking, even if all the rest of what I have ever imagined had really existed, I should have no reason for thinking that I had existed. From this I knew that I was a substance the whole essence or nature of which is to think, and that for its existence there is no need for any place, nor does it depend on any material thing; so that this "me," that is to say, the soul by which I am what I am, is entirely distinct from body, and is even more easy to know than is the latter; and even if the body were not, the soul would not cease to be what it is.

He then described how his mind conceived clearly and distinctly of an all-perfect being, and since for it not to exist would be an imperfection in it, it had to exist: The existence of the perfect being was implied in the idea of God just as, he says, the fact that the sum of the angles of a triangle is 180° is implied in the idea of a triangle. The existence of God was therefore as certain as the results of mathematics; there was more evidence for it than for the existence of the physical world, which may be an illusion, like something we see in a dream. In fact, instead of proving the existence of God from design in nature (which John Stuart Mill said was the only argument with possibilities), he proved that the physical world existed from the existence of God because God would not deceive us. Thus for Descartes, unlike for most philosophers, the existence of the physical world was more difficult to establish than the immortality of the soul and the existence of God, and in fact could not be established without first proving that God existed. He turned the usual order of things upside-down.

Part V begins with a review of all the theorems about the world that Descartes was able to prove using his method. The physical world that we live in, he said, obeyed laws that follow directly from the attributes of God; they are necessary, so that, in a sense, we have here the idea that this is the only possible world:

> Even if God had created other worlds, He could not have any in which these laws would fail to be observed.

The laws of nature, then, follow from the perfection of Deity, a proposition that John Stuart Mill was to attack in his *Essay on Nature*. These laws are mathematical, and any other world that God created would turn out to be exactly like this one we now have. God did not need to create the world exactly as we now see it; it would have evolved thus even if He had only produced the chaotic matter and allowed the laws to act upon it, but He did so in order to save time. Descartes then went on to treat in some detail the functioning of the human heart, asking his reader to dissect the heart and lungs of a great mammal as they proceed through his description. The section ends with an account of how the soul of a man differs from that of an animal, *viz.*, the man's has reason, something independent of body and therefore not mortal, that is, immortal.

> For next to the error of those who deny God, which I think I have already sufficiently refuted, there is none which is more effectual in leading feeble spirits from the straight path of virtue, than to imagine that the soul of the brute is of the same nature as our own, and that in consequence, after this life we have nothing to fear or to hope for, any more than the flies and the ants.

Like St. Paul, Descartes was one of those people who were obsessed with death. He just could not believe that his mind could stop thinking, any more than there could cease to be circles and triangles.

As for that reason which Descartes said distinguishes the soul of man from that of an animal, what is the sign of it? The sign of reason, according to Aristippus, the Socratic philosopher, was mathematics.

Finally, in Part VI, Descartes told how he had delayed the publication of his scientific discoveries when he heard of the condemnation of Galileo, lest any of the opinions he had expressed be found offensive by the authorities. He was tempted to change his mind when he realized that by keeping his method to himself, he was holding up the advancement of the human race, which would benefit from the truths that his procedures made it possible to discover. Should he, for the good of humanity, allow his treatise to be published, and invite all men of learning to adopt his method and communicate to him the various discoveries that they should make by using it, so that he might circulate them to all? Indeed, he hoped for significant discoveries, especially in medicine, which he considered the only real hope for the improvement of the human condition:

> The mind depends so much on the temperament and dispositions of the bodily organs that, if it is possible to find a means of rendering men wiser and cleverer than they have hitherto been, I believe that it is in medicine that it must be sought.

No, Descartes finally decided that he should not go public because 1) the inevitable controversies that his writings would arouse would disturb the peace and quiet he required for further progress, 2) the contributions of others would probably be full of mistakes and superfluities, and 3) there is no better way of ensuring progress in science than to let the individual genius alone, and encourage him by protecting his precious leisure from the importunities of others. Nevertheless, as a sort of compromise, he relented and published three scientific appendices, on meteors, on optics, and on geometry because 1) he did need to interest other scientists in helping him with necessary experiments and 2) he did not want to make people think that he was keeping quiet because he had something criminal to hide.

catastrophe The Greek noun καταστροφή means *a turning upside down*, from the verb καταστρέφω, *to turn* (στρέφω) *upside down* (κατά). Like *chaos*, it was chosen as a mathematical technical term for the purpose of popularization.

category The Greek word κατηγορία means *accusation*. It is a formal charge leveled in the assembly (ἀγορά) against (κατά) an opponent. Aristotle used the noun for one of the ten sets into which he divided all things in heaven and on earth. The ten categories are:

GREEK	LATIN	ENGLISH
1. οὐσία (being)	*substantia*	substance
2. πόσον (how much?)	*quantum*	quantity
3. ποῖον (how?)	*quale*	quality
4. πρὸς τί (in what way?)	*relatio*	relation
5. ποῦ (where?)	*locus*	place
6. πότε (when?)	*tempus*	time
7. κεῖσθαι (to lie)	*situs*	position
8. ἔχειν (to have)	*habitus*	possession
9. ποιεῖν (to do)	*actus*	activity
10. πάσχειν (to have done to one)	*passio*	passivity

Those looking for secondary sources on the philosophy of Aristotle may profitably consult *Aristotle* by A. E. Taylor, revised edition, Dover Publications, New York, 1955, and the chapter "Aristotle" in Will Durant's *The Story of Philosophy*, Simon and Schuster, New York, 1927, a book that made its author a millionaire.

catenary *Catena* is the Latin word for *chain*. By adding the adjectival suffix *-arius* to the stem, one produces the adjective *catenarius, -a, -um*, which means *pertaining to the chain*. From this one produced the

English noun *catenary*, the curve resembling a hanging chain. *Catenary* is the name for the curve of suspension of a flexible and inelastic chain of constant density. Galileo (1564–1642) believed that the catenary was a parabola, but Huygens showed that not only was it not a parabola, but it was not even an algebraic curve. Jakob Bernoulli, after the development of calculus, posed once again the problem of determining what the catenary was, and the correct solution was given by Leibniz (1646–1716), that it is the hyperbolic cosine. Galileo's conjecture, however, is correct if one assumes that the chain is of constant horizontal density rather than of constant density.

catenoid This is the weird combination of the Latin *catena*, *chain*, and the Greek εἶδος, *shape*. It is the surface of revolution produced by revolving a *catenary* about its axis of symmetry.

caustic This word comes from the Latin adjective *causticus*, which itself is the transliteration of the Greek adjective καυστικός, which means *burning*, from καίω, *to burn*.

cell The Latin noun *cella* means a *room, hut*.

centenarian This is a fellow who has lived to the age of one hundred years. It is derived from the Latin adjective *centenarius*, which means *related to one hundred*, for *centum* means *one hundred*. The *CBS Evening News* uses the absurd substitute *superager* for this word, a result of the lack of even the most elementary familiarity with Latin. Since *superager* does not convey the information afforded by *centenarian*, *viz.*, that the person in question has reached one hundred, it is an example of that class of modern words that are not as informative as the words they regrettably replace.

center This is the Greek noun κέντρον, *a point, prickle, spike, spur*, used by Euclid for the point around which a circle is described. It came into Latin as *centrum*, that is, by simple replacement of the Greek second-declension neuter nominative singular ending by the corresponding Latin ending.

centesimal *Centesimus* means *hundredth* in Latin, from *centum, one hundred.* Someone foolishly added the adjectival ending *-al* to what was already an adjective to produce the English adjective *centesimal.*

central tendency The later Latin authors treated *centrum* as a Latin word and made the adjective *centralis* from it by adding the Latin adjectival suffix *-alis*, whence evolved our *central.* This is a term used to describe the tendency of data to group itself around a mean, median, or mode. Since the mean requires more skill to find, the median and mode are used by less expert people. The central limit theorem of probability, so-called because of its importance, says that the distribution of the standardized mean of a random sample of any population with moment-generating function tends toward the standard normal distribution as the sample size goes to infinity.

centralizer This word was produced by adding a Latin nominal suffix of agent to a Greek verbal suffix to a Latin adjectival suffix attached to the stem of a Greek noun. Words formed by the concatenation of suffixes in this manner are ugly.

century A company of one hundred men was called a *centuria* in ancient Rome. In English the word is now commonly applied only to a term of one hundred years.

centroid This word is concocted from κέντρον, *center*, and εἶδος, *shape*. It should mean *shaped like a center*, a nonsensical description. It is an unintelligently concocted word invented in the nineteenth century for the center of mass of a homogeneous rod, lamina, or solid.

chance The Latin verb *cado, cadere, cecidi, casus* means *to fall.* From this verb proceeded the Old French *cheoir* with the same meaning, and from that verb the noun *chance* with the meaning *a fall [of the dice].*

chaos The Greek noun τὸ χάος means *the wide empty space, the gulf, the chasm.* Its choice as a technical term in the theory of dynamical systems was a great marketing success.

character The Greek noun χαρακτήρ means *a fellow who mints coins, that which is cut in or marked, the impress on stamps or coins,* from the verb χαράσσω, *to sharpen or engrave.*

characteristic From the noun χαρακτήρ there developed the adjective χαρακτηριστικός with the meaning *typical,* which came into English after the adjectival ending -ός was dropped.

chemicograph This ugly word is the name of those graphs that illustrate how atoms are bound to one another to form molecules. They are discussed by Finkbeiner and Lindstrom on pages 222–223 of *A Primer of Discrete Mathematics,* W. H. Freeman and Company, New York, 1987. *Chemicograph* is the clumsy combination of *chemic* and *graph.* The Greek noun χημεία means *the transmutation of metals.* Of this the Arabs made الكيميا , which came into Europe as *alchemia,* the origin of our word *alchemy.* The prefix *al-,* the transliteration of the Arabic definite article, was dropped, and the adjectival suffix *-icus* of Greek origin was added to produce the modern Latin *chemicus.*

chi This is the name of the Greek letter X, χ, which was pronounced like the German *ch* in *doch.* English speakers pronounce it like *k* since its true sound is not used in their language.

chi-square This is the name of a probability distribution studied by Karl Pearson. See the entry **square.**

chord The Greek noun χορδή means *a string of gut, a string of the lyre, a string of the bow, a sausage.* It was transliterated into Latin as *chorda* and thereafter often spelled incorrectly *cord.* It was used by the Greeks in the mathematical sense of our chord. Greek trigonometry consisted entirely of the study of the relationships between chords and angles; the definition of the trigonometric functions as ratios is an event of the eighteenth century.

chromatic The Greek noun χρῶμα, χρώματος means *the surface of the body, the skin, the color of the skin, color in general.* From this noun there was derived the adjective χρωματικός, *pertaining to color.* The *chromatic* *number* of a graph is the smallest number of colors needed to color the vertices of a graph in such a way that no two adjacent vertices have the same color.

cipher This word is an attempt at the transliteration of the Arabic adjective صفر, which means *empty*.

circle The Latin word *circulus* means *a circle*. The Latin noun *circus* is the transliteration of the Greek κίρκος, which means *a ring*. The Greek word for *circle* is κύκλος, which is not related to κίρκος.

circulant The Latin verb *circulor, circulari, circulatus sum* means *to gather in a circle for conversation, to gather a group around oneself*. It is a denominative verb from the noun *circulus*. (See the previous entry.) *Circulans, circulantis* is the present participle of this verb. The *Oxford English Dictionary* quotes Burnside as defining the noun *circulant* in 1881 in his *Theory of Equations*, Chapter xi, §129:

> Here in all the rows the constituents are the same five quantities taken in circular order, a different one standing first in each row. A determinant of this kind is called a *circulant*.

The adjective *circulant* appears in the expression *circulant matrices*; these are square matrices whose i^{th} row is obtained by circularly permuting the entries of the $i - 1^{st}$ row by a shift to the right. Neither as a noun nor as an adjective is the English word *circulant* a happy choice etymologically for the phenomenon that it describes; its Latin meaning does not fit its mathematical meaning.

circumcenter The Latin preposition *circum* means *around*. *Center* is just the Greek noun κέντρον, *a point, prickle, spike, spur*, used by Euclid for the point around which a circle is described. It was taken over into classical Latin as *centrum*, so the word *circumcenter* is not macaronic. The *circumcenter* of a triangle is the center of the circumscribed circle.

circumference The Latin preposition *circum* means *around*. The Latin verb *fero, ferre, tuli, latus* means *to carry*. Thus, *circumfero* is a good Latin word for *to carry around*. Its present participle is *circumferens, circumferentis*, whence one gets the noun *circumferentia* and the English *circumference*.

circumscribe The Latin preposition *circum* means *around*. The Latin verb *scribo, scribere, scripsi, scriptus* means *to write*. Thus, *circumscribo* is a good Latin word for *to write around*.

cissoid This word is composed of the two Greek nouns κισσός, *ivy*, and εἶδος, *shape*. It is the name of the curve defined by Diocles (*circa* 200 B.C.) which is supposed to look like a strand of hanging ivy. It was used by that mathematician to draw a line of length $2^{1/3}$, that is, to double the cube; see, for example, Smith's *History of Mathematics*, vol. II, pages 314–315. Let C be the circle in the polar plane with center $(a/2, 0)$ and radius $a/2$. Let ℓ be the line with equation $r = a \sec \theta$. For every angle θ, $-\pi/2 < \theta < \pi/2$, the ray m through the origin that makes an angle θ with the initial line will intersect C at some point C and will intersect ℓ at some point D. The *cissoid* is the locus of the point P on m defined by $OP = OD - OC$. Its equation is $r = a \tan \theta \sin \theta$. The line ℓ is an asymptote, and there is a cusp at the pole. Newton was able to construct it mechanically with a T-square. Fermat proved in 1661 that the area of the plane region between the cissoid and its asymptote is $3\pi\, a^2/4$. Huygens, in 1657, was able to rectify it, that is to say, to find the length of a finite piece of it. The cissoid is the pedal curve of the parabola with respect to its vertex. The definition of the cissoid may be generalized by allowing C and ℓ to be any two curves whatsoever with the property that every line through a given point Q intersects C and ℓ once each, at points P_1 and P_2, respectively. The cissoid of C and ℓ with respect to Q is then defined to be the locus of all points P such that $QP = QP_2 - Q_1P_1 = P_2P_1$.

class The Greek noun κλῆσις means *a calling, a summons, a name*, and is derived from the verb καλέω, *to call*. The related Latin noun *classis*,

classis means *one of the divisions into which Servius Tullius, the sixth legendary king of Rome, divided the people*, and later, *the fleet*, in the sense of the whole body of citizens assembled in the naval forces. Its mathematical use is the same as that of *set* or *family*.

classical See the preceding entry for the Latin noun *classis*. The adjective corresponding to the noun *classis* is *classicus, -a, -um*, not *classicalis, -a, -um*. The correct English adjective should therefore be *classic*. The additional Latin suffix *-alis* was added on top of what was already an adjective by people who no longer felt the adjectival force of *classic*. It is the same phenomenon that produced *mathematical*. The adjective *classicus* meant *belonging to a class*, and then, by specialization, *belonging to the first class*.

classics, the A theorem, paper, or book is a *classic* if it has deserved that preëminence among other theorems, papers, or books that the Greek and Latin languages and literatures were at one time held to enjoy among other languages and literatures; those products of the human mind are *classical* that enjoy the authority and respect due to that which is sublime.

According to Gibbon, there are three reasons to study the classics:

> But the Moslems deprived themselves of the principal benefits
> of a familiar intercourse with Greece and Rome, the knowledge
> of antiquity, the purity of taste, and the freedom of thought.
> (*History of the Decline and Fall of the Roman Empire*, vol. V, p. 430 of
> the first edition, 1788)

As mathematics was the creation, or rather the discovery, of the Greeks, and as its most famous book is a masterpiece of the Greek language, there is a natural affinity between it and the languages of Greece and Rome, in which the affairs of mankind were conducted for more than two millennia.

clepsydra This word is the transliteration of the Greek κλεψύδρα, from κλέπτω, *to steal*, and ὕδωρ, *water*. The *clepsydra* was a Greek water-clock used to time speeches in the courts of law. The nearest we have to it nowadays is the toy salt timer used by some people to ensure that their eggs are not boiled too long. The *clepsydra* consisted of a solid tank full of water with a spout at the bottom. As time passed, the water flowed out of the spout. For the sake of stability, the tank was in the shape of a solid of revolution. For facility of calibration, it was desirable that the water level decrease at a constant rate. It is an exercise of the application of Torricelli's law that the clepsydra is produced by the revolution of a fourth-degree equation about the vertical axis.

clockwise The late Latin noun *clocca* was taken from the Germanic languages and originally meant *bell*. Proper Latin used the word *campana* for that object. The idea is that the time is announced by the ringing of a bell. The suffix *-wise* is Germanic in origin and derived from the same root as the verb *wissen, to know*.

closed This English adjective, the second principal part of the verb *close*, is derived from the Latin *clausus*, the fourth principal part of the verb *claudo, claudere, clausi, clausus, to close*.

closure The Latin noun *clausura, a closing*, is derived from the verb *claudo, claudere, clausi, clausus, to close*.

clothoid The Greek verb κλώθω means *to twist or to spin*, and the noun εἶδος means *shape*. From the verb came the name of the Fate Κλωθώ, *Clotho*, who spun the web of life. The *clothoid* is a spiral which, according to Lawrence, was discussed by Euler in 1744 and is more commonly known as *Euler's spiral*.

co- The Latin preposition *cum* means *together*. When used as a prefix, *cum* often becomes *com-*, and the *m* is then assimilated to the following letter. To add *co-* to a noun in order to give that noun the sense of *associated* or *joint* is not Latin but English cant and is a mark of low style. In the modern Latin of the later mathematicians, such as Euler, one does find the words *cosinus, cotangens,* and *cosecans*, so these words are untouchable, but the *co-* there comes from the first two letters of *complementarius*. (See the entry **cosine**.) Already in the sixteenth century, Viète had coined the word *coefficiens, coefficient,* entirely according to the Latin rules to be found below in the entry **cum.**

cobordism The English word *border,* the French noun *le bord* with the meaning *edge,* the French verb *border, to border,* and the late Latin noun *bordatura, an edging,* and *bordus, an edge or side,* are all of Teutonic, that is, of barbarian, origin. The word *cobordism* is the combination of the prefix *co-* (see preceding entry), *bord,* and the Greek suffix *-ism, q.v.* It is a macaronic word *par excellence.* It means the study of two sets of points of the same dimension whose disjoint union is the boundary of some set of the next higher dimension.

cochleoid The Greek κόχλος (or κοχλίας) was *a shell-fish with a spiral shell,* and then the shell itself. εἶδος means *shape.* The *e* in the middle of the word is a mistake; the correct spelling, which no one uses, is *cochloid.* It is the name of a shell-shaped spiral of Jakob Bernoulli that winds through the pole rather than around it. The equation of the spiral is $r = (a \sin \theta)/\theta$, where *a* is a constant.

code The Latin noun *codex, codicis* means *the trunk of a tree;* it later came to mean *a wooden tablet, a book, or a manuscript.*

coefficient The Latin verb *efficio, efficere, effeci, effectus* means *to do, produce, make.* The English adjective *efficient* is derived from its present active participle *efficiens, efficientis.* The addition of the prefix *co-* marks this word as a Latin construction of the later mathematicians, in this case, of François Viète (1540–1603), who wrote in Latin and used *coefficiens* in the modern mathematical sense.

cofactor *Factor* is the Latin noun meaning *maker* from the verb *facio, facere, feci, factus, to make.* The subsequent addition of the prefix *co-* marks this word as a Latin construction of the modern mathematicians. The use of the word *factor* in the mathematical sense is traced by the *Oxford English Dictionary* back to 1673, where the definition appears in Kersey's *Algebra.* The definition of *cofactor* is found in an outstanding modern text:

Given the matrix $A = (a_{ij})$, let A_{ij} be the matrix obtained from A by removing the ith row and jth column. Let $M_{ij} = (-1)^{i+j} \det A_{ij}$. M_{ij} is called the *cofactor* of a_{ij}. Prove that

$$\det A = a_{i1} M_{i1} + \cdots + a_{in}M_{in}.$$

(Problem 5 on p. 292 of Herstein, *Topics of Algebra*, Blaisdell Publishing Company, Waltham, Massachusetts, 1964)

cofunction This is a low word. The deponent verb *fungor, fungi, functus sum* means *to occupy oneself with anything, to perform, to discharge [an office].* From its fourth principal part there was formed the noun *functio, functionis, that which is performed or discharged*, by the addition of the nominal ending *-io*. The English noun *function* comes from the stem of the Latin parent. The prefix *co-* was then added to produce the modern word *cofunction*. The sine and cosine, the tangent and cotangent, and the secant and cosecant, are called *cofunctions* of one another. The notion may be generalized as follows: two functions f and g are *cofunctions* if $f(\pi/2 - x) = g(x)$ and $f(x) = g(\pi/2 - x)$ wherever both f and g are defined.

Cogito ergo sum This is the Latin translation of the first result of the system of Descartes stated in Part III of the *Discourse of Method*; it means *I think, therefore I am*. It is the most famous sentence in philosophy.

cogredient The Latin verb *gradior, gradi, gressus sum* means *to step or walk*. The verb *congredior, congredi, congressus sum* means *to meet*, but it is not the ancestor of *cogredient*, which is a modern invention identifiable by the prefix *co-*. If the vector $\mathbf{x}$ is the column vector in the vector space Υ whose i^{th} entry is ξ_i, and $\mathbf{x}'$ is the row vector in the dual space Υ' whose i^{th} entry is ξ_i, then the vectors $\mathbf{x}$ are said to vary *cogrediently*. See the entry **contragredient** and Halmos, *Finite Dimensional Vector Spaces*, Springer-Verlag, New York, 1974, page 83. For the difference between *gradient* and *gredient*, see the entry **gradient**.

cohomology This is a bad word, the corruption of the first two letters of a Latin prefix having been added to a word of Greek origin. See the entries **co-** and **homology**.

cohorts, communities A headline on page 1 of the fall 2010 *Newsletter* of the American Mathematical Society reads "2010 Mathematics Research Communities." We read, "The MRC summer conferences, funded by the National Science Foundation, creates research cohorts of young mathematicians that, over time, foster joint research and coherent research programs reaching into all areas of mathematics." It is a comical highfalutin word. A similar absurdity is the use of *academy* for just plain *school*. The Latin noun *cohors* is a unit of the Roman army.

coin This is the corruption of the Latin noun *cuneus*, which means *a wedge*. It is thus related to *cuneiform*, the adjective that describes the writing of the Mesopotamians. The study of the coin toss is the central subject matter of the theory of probability.

coinitial The Latin verb *ineo, inire, inivi, initus* means *to go in, to enter*, and from it is derived the noun *initium* meaning *the beginning*; the adjectival suffix *-alis* is added to the stem of this noun to produce the adjective *initialis*, from whence comes our word *initial*. The prefix *co-* has been added to produce the modern English word *coinitial*. If all the vectors in a vector space are imagined to issue from the same point of origin, they are called *coinitial*.

coincident The Latin verb *incido, incidere, incidi, incasus* means *to fall on*. The adjective *incident* is derived from its present active participle *incidens, incidentis*. The prefix *co-* has been added to produce the modern (seventeenth century) English word *coincident*. The Greek verb *to coincide* is ἐφαρμόζω, which is composed of the verb ἁρμόζω, *to fit together*, and the preposition ἐπί, which means *on*. It is the word Euclid used for this idea. In the twelfth century, Adelard of Bath translated the verb ἐφαρμόζω by *superincido* and *supervenire*.

collect This verb is formed from the fourth principal part of the Latin verb *colligo, colligere, collegi, collectus*, which means *to bring together*. Suppose one has equal numbers of k different kinds of coupons in an urn, and one samples from the urn with replacement, each sampling consisting of drawing one coupon at random from the urn. Let **N** be the random variable whose value n is the number of drawings required until one has a complete collection of coupons for the first time. The study of the properties of **N** is called the *coupon collector problem*, a standard topic in the best probability courses. The expectation and variance of **N** are given by

$$E(N) = k(1 + 1/2 + 1/3 + \cdots + 1/k)$$

$$Var(N) = k^2(1 + 1/2^2 + 1/3^2 + 1/4^2 + \cdots + 1/k^2) -$$

$$k(1 + 1/2 + 1/3 + \cdots + 1/k)$$

collection This word is derived from the Latin noun *collectio, collectionis*, which comes from the verb *colligo* mentioned in the preceding entry.

collinear There is a Latin verb *collineo, collineare, collineavi, collineatus* that Cicero used with the meaning *to direct something in a straight line* (*linea*), but the English word is of the nineteenth century. The English noun *line* is produced by dropping the nominative case ending *-a* from the Latin noun *linea*.

column This is the Latin word *columna* of the same meaning with the case ending dropped.

combination The Latin adjective *bini* means *twofold, two apiece*. The associated verb *combino, combinare, combinavi, combinatus* means *to list two and two*. The number of combinations (different subsets) of size r from a set of n different elements is $n!/[r!(n-r)!]$.

combinatorial The noun of agent *combinator, combinatoris*, is *a fellow who puts things two by two*; it is formed by adding the suffix *-or* to the stem of the fourth principal part of the verb *combino*. The modern English adjective *combinatorial* was formed by adding the stem of the Latin adjectival suffix *-alis* to the adjective *combinatory*, itself a vestige perhaps of a medieval Latin adjective *combinatorius*.

combinatorics The name of the subject that deals with permutations and combinations was concocted incorrectly on the analogy of *mathematics* by adding *-ics* to the Latin noun *combinator*. However, the English ending *-ics* is the Greek suffix -ικ with the English plural *s* added; it is poor style to apply it to the end of a Latin word or a word formed on a Latin model. The word *combinatorics* has thus the dubious honor of being composed of elements of three different languages, a real mongrel. If words were formed on this analogy, we would have branches of mathematics called *annihilatorics* and *factorics*. The correct name for this subject would have been *combinatoria*.

commensurable The Latin deponent verb *commetior, commetiri, commensus sum* means *to measure against something, to compare*. The addition of the suffix *-abilis* produces the adjective *commensurabilis* that Boëthius used with the meaning *capable of comparison, having a common measure*.

common This is derived from the Latin adjective *communis* of the same meaning, itself produced from the compounding of the preposition *cum* and the root *mu-*, *to bind*.

commutative The Latin verb *commuto, commutare, commutavi, commutatus* means *to change, to change one thing for another*. The addition of the suffix *-ivus* to the verb stem produces an adjective pertaining to the verb; so, *commutativus* would mean *pertaining to exchanging*.

commutator The addition of the nominal suffix of agency *-or* to the stem of the fourth principal part of the Latin verb *commuto* produces the noun *commutator* with the meaning *one who exchanges*.

commute The English verb is derived from the first principal part of the Latin verb *commuto, to change, to change one thing for another*.

compact *Compactus* is the fourth principal part of the Latin verb *compingo, compingere, compinxi, compactus,* which means *to put together, to construct,* itself formed from the preposition *cum, with,* and the verb *pango, pangere, panxi, pactus, to fasten, fix, drive in, compose, write.*

compactification This is the stem of the make-believe Latin noun *compactificatio, compactificationis.* The suffix *-ficatio* comes from the verb *facio, facere, feci, factus, to do,* and indicates the *making* of whatever is indicated by the stem to which it is appended.

compactness The suffix *-ness* is of Germanic origin; the adjective *compact* is of Latin origin. See the preceding two entries. The word means *the quality or condition of being compact.* It was in the English language for centuries before it became a mathematical technical term.

companion of the cycloid A companion is someone with (*com-*) whom one breaks bread (*panis*). The *companion of the cycloid* is the name of a curve defined by Roberval (1602–1675). Let E be any point on the circle of radius a and center $C(a\pi, a)$. Let AB be the vertical diameter, with B on the x-axis. Let θ be the angle BCE. Drop the perpendicular DE from E to AB. Go out from the circle along the extension of DE to a point P such that DP is the length of the arc from A to E. If this is done for every point E on the circle, the points P form the *companion of the cycloid,* so-called because of its similarity to the cycloid. It is a sinusoid.

compass The Latin verb *pando, pandere, pandi, pansus* (also *passus*) means *to stretch out, extend.* From its fourth principal part is derived the fourth-declension noun *passus, passūs* with the meaning *step, stride, pace.* From this noun the common folk in later times made the verb *compasso, compassare,* which entered Italian with the meaning *to go around.*

complement The Latin noun *complementum* means *that which fills up.* It is derived from the verb *compleo.* (See the following entry.) The

English word *compliment* is derived from the same Latin word; it means *the fulfilling of the laws of courtesy*. The two different spellings were used interchangeably for some time, although the one with the *e* is etymologically correct. In modern usage, the mathematical term is *complement*, and the word *compliment* is reserved for a courteous observation about someone's appearance or station.

complete The Latin verb *compleo, complere, complevi, completus* means *to fill up*. It is compounded of the prefix *com-* and the rare verb *pleo, plere*, which means *to fill*; the force of the prefix *com-* is to intensify that meaning. The English adjective and verb are derived from the fourth principal part of *compleo*. To *complete the square* is the process of adding rectangles to a figure in order to make it a square; the name comes from the geometrical algebra of the Greeks. See Euclid's *Elements of Geometry*, Book II. The modern algebraic equivalent is the following calculation:

$$ax^2 + bx + c = a[x^2 + (b/a)x + (c/a)] =$$

$$a[x^2 + (b/a)x + (b^2/4a^2) + (c/a) - (b^2/4a^2)] =$$

$$a\{[x + (b/2a)]^2 + [(c/a) - (b^2/4a^2)]\}.$$

If we set this formula equal to zero and solve for *x*, we produce the *quadratic formula*:

$$x = [-b + (b^2 - 4ac)^{1/2}]/2a, \text{ or}$$

$$x = [-b - (b^2 - 4ac)^{1/2}]/2a.$$

complex The prefix *com-* is derived from the preposition *cum, together*. The Latin verb *complector, complecti, complexus sum* means *to braid together, to embrace, to hold fast*. The Latin language has the word *plexus, -a, -um* with the meaning *braided*, which is in form a participle as if from a verb *plecto, plectere*, but no Latin verb *plecto* exists. There is, however, a Greek verb πλέκω, πλέξω, ἔπλεξα, *to braid*. The Greek verb πλέκω is the same as the Latin *plico, plicare, plicavi, plicatus*, which means *to fold*, and which when combined with the prefix *com-* produces *complico, complicare, complicavi, complicatus* with the meaning *to fold together*, that is, *to make intricate*. Thus, the basic idea is that of something constructed of parts and therefore possibly complicated. The complex numbers have two parts, a real part and an imaginary part.

component The Latin verb *compono, componere, composui, compositus* means *to put (pono) together (cum)*. This adjective is derived from the stem of its present participle *componens, componentis*, which has the meaning *putting together*.

composite This adjective is derived from the fourth principal part of the Latin verb *compono, componere, composui, compositus*, which means *to put together*.

composition The Latin noun *compositio, compositionis* is formed from the fourth principal part *compositus* of the Latin verb *compono, componere*, which means *to put together, to compose*.

compound The Latin verb *compono, componere, composui, compositus* entered Old French as *componre* and then *compondre*, whence is derived our *compound*.

compute The Latin verb *computo, computare, computavi, computatus* means *to recken (puto) together (com-), to calculate*.

computer The word *computer* is derived from the Latin verb *computo, computare*, which means *to reckon (puto) together (com- from cum)*, *to calculate, to show*. The English suffix *-er* has been added to the stem of the verb to indicate agency, the thing doing the computing.

concatenation The concatenation of digits is used to indicate permutations. For example, 00110101011 is a permutation of length eleven of digits taken from the set $\{0,1\}$. The word conveys the notion of linkage, which is suggested by the Latin *catena (chain)* and *con-* (*with*). The late Latin verb *concatenare* means *to link or bind together*. The verb *to concatenate* is good English.

concave The Latin adjective *concavus* means *hollow, vaulted, arched*. The prefix *con-* (from *cum*) intensifies the meaning of the adjective *cavus*, *hollow*.

concavity The addition of the nominal suffix *-itas* to the stem of the adjective *concavus* produced the noun *concavitas*, which entered French as *concavité*, which became *concavity* in English.

concentric This is a low word. The Latin prefix *con-* from the preposition *cum* has been added to the stem of the Greek adjective κεντρικός, *pertaining to the center* (κέντρον). The proper word would have been *syncentric*, formed entirely from Greek elements, or *concentral*, composed entirely of Latin parts.

conchoid The Greek noun κογχή means *mussel or shell*, and εἶδος means *shape*. The conchoid is therefore a *shell-shaped curve*. It is a curve first studied by Nicomedes (*circa* 225 B.C.), which, if allowed, permits the trisection of an arbitrary angle. Let C be the line in the polar plane whose equation is $r = a \sec \theta$, and let the positive parameter b be given. On each line ℓ through the origin, mark off the points P and P' that are a distance b on each side of C from the point of intersection of ℓ and C. Then the locus of P and P' is the *conchoid of Nicomedes*. This curve has two branches, one on each side of the asymptote C. If $a = b$, there is a cusp at the pole. If $a < b$, there is a loop through the pole. The definition of Nicomedes was generalized by Pascal, who allowed C to be any curve whatsoever that is intersected by any ray through the origin at one point, not counting the origin itself if the origin is on C. If C is a circle with center at the origin, then the conchoid is a *limaçon*.

conclude The Latin verb *concludo, concludere, conclusi, conclusus* means *to shut or close* (*claudo*) *up* (*con-*), *confine*, and eventually, as a philosophical technical term, *to bring an [argument] to its logical end*. The prefix *con-* here intensifies the force of the verb.

conclusion The Latin noun *conclusio, conclusionis* with the meaning *a shutting, a closing, a peroration, last step of a syllogism*, came into French as *conclusion*, and from thence into English.

concrete The Latin verb *concresco, concrescere, concrevi, concretus* means *to grow, to become stiff*. It is composed of the prefix *com-* from the preposition *cum* (*together*) and the fourth principal part of the verb *cresco, crescere, crevi, cretus, to grow*.

concurrent This adjective comes from the present participle *concurrens, concurrentis* of the Latin verb *concurro, concurrere, concurri, concursus*, which means *to run (curro) together (con-)*.

condition The Latin noun *conditio, conditionis* is related to the verb *condico, condicere, condixi, condictus,* which means *to come to an agreement with, to agree to, to fix, to settle*. The noun means *an arrangement, stipulation*. It came into French and then English as *condition*.

conditional The Latin adjective *conditionalis*, with the meaning *pertaining to a condition*, was formed by adding the adjectival suffix *-alis* to the stem of the noun *conditio, conditionis*. The English word is just the stem of the Latin adjective. A *conditional inequality* is one that is not true for all values of the variable, for example, $x^2 > x$.

cone The Latin noun *conus* is the transliteration of the Greek noun κῶνος.

confidence [interval] The Latin verb *fido, fidere, fisus sum* means *to trust*. The compound verb *confido, confidere, confisus* means *to have complete trust*. The prefix *con-* here simply strengthens the verbal element, as often. Its present participle is *confidens, confidentis*, from whence came the noun *confidentia, complete trust*. This noun came into French as *confidence* and was then taken over into English. See also the entry **interval**.

configuration The Latin verb *fingo, fingere, finxi, fictus* means *to shape, fashion, or form*. Our word *fiction* comes from the fourth principal part. The Latin noun *figura* proceeded from it, and from the noun was produced the derived verb *figuro, figurare, figuravi, figuratus*, also with the meaning *to shape, fashion, or form*. The addition of the prefix *con-* produced the later verb *configuro*, from whence came the noun *configuratio, configurationis*, from whose stem proceeded the French and English noun *configuration*.

confocal This is a modern formation, the nineteenth-century offspring of uniting the adjectival suffix *-alis* to the union of *con* (from *cum, together*) and *focus* (*fireplace*) as if to produce a Latin adjective *confocalis* with the meaning *having the same foci or focus*.

conformal The Latin noun *forma* means *shape, form*; by adding the suffix *-alis* to the stem there was formed the adjective *formalis* with the meaning *pertaining to form*. The ecclesiastical Latinists then formed the adjective *conformalis* by adding the prefix *con-* to the stem. The resulting adjective means *conformable*, and its stem is the English adjective. There is also the related verb *conformo, conformare*, which means *to join, adapt, or form together*.

congruence This noun is the French and English transformation of the Latin noun *congruentia*, which means *a running together, a meeting*. See the following entry.

congruent The Latin verb *congruo, congruere, congrui* means *to run together, to come together, to meet*. Lewis and Short say that its etymology is uncertain. From the stem of its present participle *congruens, congruentis* was formed the English adjective *congruent*. Euclid did not have a special word for *congruent*; he just spoke of *equal* triangles.

conic The Greek noun κῶνος means *the pine tree*, and from it was formed the adjective κωνικός, *pertaining to the pine tree*. From this adjective proceeds the English adjective *conic*. The form **conical** is a mistake, the result of superimposing the Latin adjectival suffix *-alis* upon what was already an adjective in Greek. Such Greek adjectives were sometimes erroneously imagined to be nouns, and a further measure of ignorance caused the Latin ending to be appended.

conical See the preceding entry.

conjecture The Latin verb *conicio, conicere, conieci, coniectus* means *to throw together* from the prefix *con-* (*together*) and the verb *iacio* (*throw*). The Latin noun *coniectura* was formed from the future active participle *coniecturus, -a, -um* and has the meaning *a guess*. The addition of the nominal suffix of agency *-or* to the stem produces the noun *coniector*, an interpreter of guesses, riddles, and dreams. The addition of the feminine suffix *-trix* produces *coniectrix*, a female interpreter of dreams. In the formation of English nouns, the feminine suffix is rarely tolerated became of a fear that it injects an element of comedy and detracts from the dignity of the female agent; instead, the masculine form is preferred. For example, a woman who is a teacher is called an *instructor, professor,* or *doctor* just like the man, and not an *instructrix, professtrix,* or *doctrix,* which would be correct Latin.

conjugacy The Latin noun *coniugatio* means *the quality of being yoked together*. It is derived from the verb *coniugo*. (See the following entry.) The *j* is simply the letter *i* in its capacity as a consonant instead of a vowel.

conjugate The Latin verb *coniugo, coniugare, coniugavi, coniugatus* means *to yoke* (*iugo*) *together* (*con-*). From its fourth principal part was formed the English verb *to conjugate* and the adjective *conjugate*. The Latin adjective *coniugalis* means *pertaining to the marriage bond*.

conjunction The Latin verb *coniungo, coniungere, coniunxi, coniunctus* means *to join together*. From its fourth principal part was formed the noun *coniunctio, coniunctionis*, from whose stem was produced the French and English noun *conjunction*.

connected The late Latin verb *conecto, conectere, conexui, conexus* means *to bind together*. It is composed of the prefix *cō-* (derived from the preposition *cum, with*) and the verb *necto, nectere, nexui, nexus*, which means *to bind.* The spelling *cōnecto* is classical Latin; the later spelling *connecto* is less correct. The English verb *connect* was formed from a late form of the noun *conexio, conexionis*, after *cō-* had become *con-* and the *x* had been changed to *ct. Conexio* was formed from the fourth principal part of the verb *conecto* in the usual manner.

conoid The Greek noun κῶνος means *the pine cone tree,* and εἶδος means *shape.* The English *conoid* is the metamorphosis of the Greek adjective κωνοειδής, which means *cone-shaped.*

consecutive The Latin adjectival suffix *-ivus* was added to the stem of the past participle *consecutus* of the verb *consequor* to form the adjective *consecutivus.* See the following entry.

consequence The Latin verb *consequor, consequi, consecutus* means *to follow (sequor) along with (con-).* Its present participle *consequens, consequentis* means *following after, logically appropriate.*

conservative The Latin verb *conservo, conservare, conservavi, conservatus* means *to keep, to preserve* and is itself derived from *cum, together,* and *servo, servare, to watch over, keep.* The adjectival suffix *-ivus* was added to the stem of the fourth principal part. If the word *conservativus* ever existed, it was in medieval Latin.

consistent The Latin verb *sisto, sistere, stiti, status* means *to stand* or *to cause to stand.* Strengthening the verb by the addition of the prefix *con-* from *cum* results in the verb *consisto, consistere, constiti, constitus,* which means *to put oneself in any place, to agree with, to be formed of, to stand still.* The English adjective is derived from the stem of the present participle *consistens, consistentis.*

constant This adjective is derived from the present participle *constans, constantis* of the Latin verb *consto, constare, to stand (sto, stare)* with (*con-* from *cum*), *to agree.* It also means *to stand firm,* the prefix *con-* in this case strengthening the following verb.

constraint This word is derived via the Old French *constreindre* from the Latin *constringo, constringere, constrinxi, constrictus*, which means *to draw or bind* (*stringo*) *together* (*con-*), *to confine*.

construct The Latin verb *construo, construere, construxi, constructus* means *to heap up together, to build up, to arrange*. This verb is itself composed of the prefix *con-* (from *cum, together*) and the verb *struo, struere, struxi, structus* (related to the verb *sterno, sternere, stravi, stratus* and the Greek στορέννυμι, *to stretch out*).

constructible In the Euclidean plane geometry, something existed only if it could be constructed with unmarked straightedge and compass. Such an angle, line segment, or figure was by later authorities said to be *constructible*. Plane curves produced by the use of other aids were called *mechanical* and afforded less respect.

> In 1672, the Danish mathematician Georg Mohr (1640–1697) published an unusual book entitled *Euclides danicus* in which he showed that any pointwise construction that can be performed with compasses and straightedge (that is, any "plane" problem) can be carried out with compasses alone. Despite all the insistence by Pappus, Descartes, and others on the principle of parsimony, many of the classical constructions were shown by Mohr to have violated the principle through the use of two instruments where one would suffice! Obviously one cannot draw a straight line with compasses; but if one regards the line as known whenever two distinct points on it are known, then the use of a straightedge in Euclidean geometry is superfluous. So little attention did mathematicians of the time pay to this amazing discovery that geometry using compasses only, without the straightedge, bears the name not of Mohr but of Mascheroni, who rediscovered the principle 125 years later. Mohr's book disappeared so thoroughly that not until 1928, when a copy was accidentally found by a mathematician browsing in a Copenhagen bookstore, did it become known that Mascheroni had been anticipated in proving the supererogation of the straightedge. (Carl Boyer, *A History of Mathematics*, John Wiley & Sons, Inc., New York, 1968, pp. 405–406)

contact This is the stem of the fourth principal part of the verb *contingo*. See the entry for **contingent** below.

content The Latin verb *teneo, tenere, tenui, tentus* means *to hold*. The addition of the prefix *con-* from *cum* (*with*) produces another verb *contineo, continere, continui, contentus* with the meaning *to hold together*. Our noun *content* is derived from the fourth principal part *contentus* of this verb. See the entries for *continuity* and *continuous* below.

contingency This noun was formed from the present participle *contingens, contingentis* of the verb *contingo*. The noun *contingentia* developed from the stem of the participle, and the ending *-ia* usually became *-y* in English. See the following entry.

contingent The Latin verb *tango, tangere, tetigi, tactus* means *to touch*, and the addition of the prefix *con-* from *cum* (*with*) produces the compound verb *contingo, contingere, contigi, contactus* meaning *to touch, to touch with, to happen*. The English word is the stem of the present participle *contingens, contingentis*, which means *touching, happening*.

continuity The Latin noun *continuitas, continuitatis* means *unbroken succession*. It is derived from the same Latin verb *contineo* as the noun *content, q.v.* Latin nouns ending in *-tas* saw their ending modified to *-té* upon admission into French, and to *-tà* upon their entry into Italian. The ending *-té* eventually became *-ty* in the development of English. See the following entry.

continuous The Latin adjective *continuus* means *connected up, hanging or holding together*. It is derived from the verb *contineo, continere, continui, contentus*, which means *to hold* (*teneo*) *together* (*con-* from *cum*).

continuum This word is the neuter singular of the Latin adjective *continuus, -a, -um*, which means *connected up, hanging together*. See the previous entry.

contour The rare Latin word *turnus* is *a carpenter's tool used to draw a circle, a lathe*. It is the same as the Greek τόρνος. It is the origin of the late Latin verb *torno, tornare*, which means *to turn*. When preceded by the prefix *con-* it became *contorno, contornare,* from which was developed the French noun *contour.*

contra This is the Latin preposition meaning *against.* The corresponding Greek preposition is ἀντί.

contraction The Latin verb *contraho, contrahere, contraxi, contractus* means *to drag (traho) together (con-* from *cum).* The noun *contractio, contractionis* was formed from the fourth principal part of the verb and means *a drawing together.*

contradiction The Latin verb *contradico, contradicere, contradixi, contradictus* means *to speak (dico) against (contra).* From its fourth principal part was formed the noun *contradictio, contradictionis,* from whose stem the French and English noun *contradiction* was formed.

contragredient The Latin preposition *contra* means *against,* and the verb *gradior, gradi, gressus sum* means *to step or walk.* There is no Latin verb *contragredior;* the word is a modern formation. The *Oxford English Dictionary* cites Sylvester in 1853 as the first fellow to use this adjective. If the vector $\mathbf{x}$ is the column vector in the vector space Υ whose i^{th} entry is ξ_i, and $\mathbf{x}'$ is the row vector in the dual space Υ' whose i^{th} entry is ξ_i, then the vectors $\mathbf{x}'$ are said to vary *contragrediently.* See the entry **cogredient** and Halmos, *Finite Dimensional Vector Spaces,* Springer-Verlag, New York, 1974, page 83. For the difference between *gradient* and *gredient,* see the entry **gradient.**

contraindications This strange word appears on page 379b of the article "Tribute to Vladimir Arnold" in the March 2012 issue of the *Notices of the American Mathematical Society* (vol. 59, no. 3, pp. 378–399). It is an example of a word that needs the hyphen between the parts because of its singularity. Without the hyphen, the eye expects *contradictions,* and so one halts and has to read the sentence again. There would be no such waste of time if it were spelled *contra-indications.*

contraposition The Latin verb *contrapono, contraponere, contraposui, contrapositus* means *to place opposite to, to oppose to*; it is composed of the preposition *contra, against,* and the verb *pono, ponere, posui, positus, to put.* From its fourth principal part was formed the noun *contrapositio, contrapositionis,* found in Boëthius, whose stem is the English and French noun. Its meaning is *the mode of deducing the statement $\sim B \Rightarrow \sim A$ from the statement $A \Rightarrow B$.*

contrapositive See the preceding entry. This word was formed late by adding the adjectival suffix *-ivus* to the stem of the past participle *contrapositus.* The *Oxford English Dictionary* cites no use prior to 1870. The *contrapositive* of the statement $A \Rightarrow B$ is the statement $\sim B \Rightarrow \sim A$. They are equivalent statements.

contrary The addition of the suffix *-arius* to the stem of the Latin preposition *contra* produced the adjective *contrarius* with the meaning *opposite, over against.*

contravariant This word is compounded of the preposition *contra, against,* and the stem of the present participle *varians, variantis* of the verb *vario, variare, variavi, variatus,* which means *to diversify, change.* If the vector **x** is the column vector in the vector space Υ whose i^{th} entry is ξ_i, and **x'** is the row vector in the dual space Υ' whose i^{th} entry is ξ_i, then the vectors **x'** are called *contravariant.*

converge The Latin verb *vergo, vergere, versi* is connected with *verto, vertere, verti, versus*; both mean *to turn,* and the former also has the meaning *to bend, to be inclined.* In the post-classical period, the prefix *con-* was added to produce the verb *convergo* with the meaning *to bend together.*

conversation This is faculty cant and computer lingo for *correspondence* or *debate.* ("This conversation has been moved to the trash.") The verb *conversor, conversari, conversatus sum* is a frequentative of *converto* and means *to have dealings with.* See the following entry.

converse The Latin verb *verto, vertere, versi, versus* means *to turn, to turn around*. The addition of the suffix *con-* from *cum, together*, in this case intensifies the force of the following verb and produces the compound verb *converto, convertere, converti, conversus* meaning *to turn around forcibly* or *frequently*. Our noun *converse* comes from the fourth principal part of *converto*.

conversion The Latin noun *conversio, conversionis* is derived from the verb *converto* (see the previous entry) and means *a turning around, a periodic return*.

convex The Latin adjective *convexus, -a, -um* means *arched, vaulted*, and is related to the verb *conveho, convehere, convexi, convectus*, which means *to bring (veho) together (cum), to carry into one place, to convey*. According to the *Oxford English Dictionary*, the idea is that in forming an arch, the ends are brought together.

convolution The Latin verb *convolvo, convolvere, convolvi, convolutus* means *to roll (volvo) together (con-)* or *to roll around*. The noun *convolution* is formed as if a noun of action *convolutio, convolutionis* from the fourth principal part of *convolvo*; however, there is no such word in Latin. The English noun with the meaning *a twisting together* goes back to the sixteenth century.

coordinate When the Latin preposition *cum* is appended to a word beginning with an *o*, it is contracted to *co-*. Thus, *cum + ordinatus = coordinatus*. The verb *ordino, ordinare* means *to put in order* and is derived from the noun *ordo, ordinis*, which means *order*. The Cartesian coordinate system for the plane consists of two perpendicular directed axes, the *abscissa* and the *ordinate*, the former usually drawn horizontally, the latter vertically. Their point of intersection is the *origin*. By tradition and superstition, the right-hand side of the abscissa and the top half of the ordinate are considered positive, for most people are right-handed, and the left (*sinister*) side was considered unlucky by the Romans; furthermore, Olympus is upwards and Tartarus is downwards.

Facilis descensus Averno
Noctis atque dies, patet atri janua Ditis.
Sed revocare gradus, superasque evadere ad auras,
Hoc *opus*, hic labor est. (Aeneid VI, 126–129)

The gates of hell are open night and day.
Smooth the descent, and easy is the way.
But to return and view the cheerful skies,
In this *the task* and mighty labor lies. (Dryden's Vergil)

Distances are marked off on each axis, the same unit of measurement usually being used. There is a one-to-one correspondence between the points in the plane and the ordered pairs of real numbers *(x,y)*, where *x* is the directed distance of the point from the origin along the abscissa, and *y* is the directed distance of the point from the origin along the ordinate. Struik has concluded (p. 154) that Leibniz was the first to speak of *coordinati* in the modern sense.

coplanar *Planus, -a, -um* means *flat* in Latin. The addition of *co-* to the adjective *planar* is English; had the Romans wanted to add *cum* to their adjective *planaris*, they would have written *complanaris*, and we would have had the adjective *complanar*. Nevertheless, the use of such words as *cotangens* and *cosecans* by Euler and other Latin-writing mathematicians of the seventeenth century throws the mantle of legitimacy over this form.

coprime This adjective is a synonym of *relatively prime*. It is composed of the modern particle *co-* and the adjective *prime* from *primus, -a, -um*, which means *first*. It is a low word and the alternative *relatively prime* should be preferred.

corollary The Latin word is *corollarium*, a small crown (*corona*) placed in proximity to a larger one, for example, in paintings of royal parents and their children, where a child has the small crown and a parent the full-sized one. In mathematics, it came to denote a minor result that follows immediately from a major one. The word is the medieval Latin translation for the Greek πόρισμα, used for this purpose by Euclid.

I was once surprised to hear a colleague accent the word *corollary* on the second rather than the first syllable. This leads us to consider the question, How do we decide how to accent a word correctly? The question is not one of interest solely to mathematicians.

The first principle is that we must follow established usage when it is constant and universal; custom must be our guide, and we must do nothing that attracts attention. This means that the English accentuation of a word of foreign origin is not infallibly determined by how the word was accented in its original language. For example, the word *senator* is Latin, but was pronounced by them *se-na´-tor*. In English, though, that would be a mistake; we must say *se´-na-tor* because everyone else speaking English has always and everywhere said so, from the queen of England down to the milkmaid. This example makes clear the fact that when a word is taken over from a foreign language, its correct accentuation in English may be different from that which it had in its language of origin. In those cases, however, where usage is not constant or not universal, how ought we to make a decision? The answer is that we should follow the established practice of the most learned people in the country where we live. This is the principle *cuius regio, eius religio*.

It is difficult to correct misaccentuation, the most persistent of blunders. The bungling of accentuation by people attempting to use foreign words is annually illustrated at the highest level of academia by American college presidents when they grant degrees *honóris causa*, which they comically mispronounce *hónoris causa*. Experience has taught me that all attempts to correct this error are futile.

correct The Latin verb *corrigo, corrigere, correxi, correctus* means *to set straight (rectus) or upright*. The English word is formed from the fourth principal part of the Latin verb. *Corrigo* is the compound of *cum* (which in this case intensifies the force of the following verb) and the verb *rego, regere, rexi, rectus*, which means *to keep straight, to guide, to direct, to rule*. The fourth principal part has the meaning *straight, in a straight line*.

correlation This noun was first used in a scientific context by Galton in 1888. The adjective *correlative* was common in the French of the sixteenth century and indicates that the adjective *correlativus* existed in the Latin of the period. The parts of this word are the prefix *cor-* from *cum*, *with*, and the adjective *relativus*, for which see the entry **relative**.

correspondence The Latin verb *respondeo, respondere, respondi, responsus* means *to answer*. In the Middle Ages, the prefix *cor-* (from *cum*) was added to describe the practice of writing letters back and forth and produced the verb *correspondeo, correspondere*.

cosecant The *Oxford English Dictionary* traces the Latin word *cosecans* at least as far back as the *Opus Palatinum* of Rheticus in 1596. It is an abbreviation for ***complementarii anguli* secans**, that is, *the secant of the complementary angle*. See the entry **cosine**.

coset The addition of the Latin prefix *co-* (from *cum*, *with*) to the noun *set* is an example of the indiscriminate use of *co-* . See the entry **set**.

cosh This is the standard abbreviation for the hyperbolic cosine function: $cosh\ x = (e^x + e^{x})/2$. The abbreviation stands for *cosinus hyperbolicus*. The pronunciation *cosh* is acceptable, but it would better be read *hyperbolic cosine*. The hyperbolic cosine is the **catenary**.

cosine The *Oxford English Dictionary* traces the Latin word *cosinus* at least as far back as Gunther's *Canon Triangulorum* in 1620. The Latin noun *complementum* means *that which fills up*; it is formed from the verb *compleo, complere, complevi, completus, to fill up*. (The adjective *plenus* means *full*.) From *complementum* was formed the late Latin adjective *complementarius, -a, -um* meaning *complementary*. *Cosinus* is thus the abbreviation of ***complementarii anguli* sinus**, that is, *the sine of the complementary angle*. The word cannot be explained by the process of just adding *co-* to a noun, such as in *coauthor*, *cochairman*, which is a modern invention. The *law of cosines* in a triangle is the generalization of the Pythagorean theorem. If a triangle has sides *a*, *b*, and *c* with opposite angles *a*, *β*, and *γ* respectively, then

$$c^2 = a^2 + b^2 - 2ab\ cos\ \gamma.$$

cotangent The *Oxford English Dictionary* traces the Latin word *cotangens* at least as far back as Gunther's *Canon Triangulorum* in 1620. It is an abbreviation for ***complementarii anguli tangens***, that is, *the tangent of the complementary angle*. See the entry **cosine**.

coterminal This word was intended to convey the meaning of *having the same ending*. It is composed of the prefix *co-* and the stem of the late Latin adjective *terminalis, pertaining to the end*, formed by adding the adjectival suffix *-alis* to the stem of the noun *terminus, boundary*. But the Latin word for bordering upon or *adjacent* was *conterminus*. So, it would have been preferable if the word *conterminous* had been coined and used. The deplorable invention of *coterminal* is due to the indiscriminate habit of adding *co-* to nouns and adjectives to create new nouns and adjectives, regardless of the existence of better options.

coth This is the standard abbreviation for the hyperbolic cotangent function: $\coth x = (\cosh x)/\sinh x$. The abbreviation stands for *cotangens hyperbolica*. Someone somewhere is probably pronouncing it *cŏth*, but it should be read *hyperbolic cotangent*.

countable The addition of the suffix *-abilis* to the stems of Latin verbs produces adjectives describing the ability to undergo the action of the verb. The word *countable* entered the English vocabulary no later than the fifteenth century, when it was borrowed from the French of the time. The Latin original is *computabilis* meaning *capable of being counted*. The idea intended by the creator of this word would have been better transmitted by the word *computable*. The mistake can no longer be corrected, as *countable* has taken on a technical meaning in set theory that protects it from molestation.

count This word is derived from the Latin verb *computo, computare*, which means *to reckon (puto) together (com-* from *cum), calculate, show*.

counterclockwise This combination of the Latin preposition *contra* and the English adjective *clockwise* is a praiseworthy creation of the nineteenth century that gets the intended idea across. The late Latin word *clocca* means *bell*, and its root is the origin of *clock*. *Wise* is a derivative of the German *wissen*, *to know*. *Clockwise*, therefore, is a hybrid that would once have been repugnant to learned ears.

counterexample This modern word is the combination of *contra*, *against*, and *exemplum*, *example*. (See those entries.) It is intended to convey the meaning of an example that proves a statement wrong. It is built on the analogy of the good word *contradiction* from the Latin *contradictio*, *contradictionis*, *a speaking against*.

coupon The Greek noun κόλαφος is *a buffet, a box on the ear*. This came into the Latin New Testament as *colaphus*, which is the origin of the French verb *couper*, which means *to strike, to cut*, whence came the noun *coupon*, which entered English. A *coupon* is a piece of paper or ticket cut off from a sheet or roll. See the entry **collect** for the *coupon collector's problem*.

covariance This modern noun is composed of the prefix *co-* from *cum*, *with*, and the noun *variance*. See the entry **variance** below.

covariant This noun was invented by Sylvester. If the vector **x** is the column vector in the vector space $\mathcal{V}$ whose i^{th} entry is ξ_i, and **x'** is the row vector in the dual space $\mathcal{V}'$ whose i^{th} entry is ξ_i, then the vectors **x** are called *covariant*.

cover The Latin verb *coöperio, coöperire, coöperui, cöopertus* means *to cover* (*operio*) *entirely* (*co-* from *cum*, *together*). The force of the prefix here is to intensify the following verb. The Latin word suffered major corruption when it entered Italian as *coprire* and French as *couvrir*, from which latter the English word proceeded. If $\{X, \mathcal{T}\}$ is a topological space, a class $\{G_\alpha\}$ of open subsets of X is an *open cover* for X if for all $x \in X$ there exists a $G_x \in \{G_a\}$ such that $x \in G_x$.

coversine See the entry **haversine**.

criterion This is the transliteration of the Greek noun κριτέριον, *a consideration taken into account in making a judgment*, from κρίνω, *to judge*.

cross This word is the corruption of the Latin *crux, crucis,* which displaced the Germanic word *rood* of the same meaning.

cryptography This is the name for the mathematical science of code-breaking. The Greek verb κρύπτω means *to hide,* and the ending *-graphy* comes from the noun γραφή, which means *writing,* from the verb γράφω, *to write.*

csch This is the standard abbreviation for the hyperbolic cosecant function: *csch* $x = 1/sinh$ x. The abbreviation stands for *cosecans hyperbolica.* It should be read *hyperbolic cosecant.*

cube The Greek noun κύβος means *a die for play.* It is the same mathematical figure as the ἑξάεδρον, the *hexahedron.* The expression *to cube* is a shortened form of the Latin *deducere ad cubum,* used, for example, by Cardano.

cubic, cubical From the Greek noun κύβος meaning *a die for play* was formed the adjective κυβικός with the meaning *pertaining to a die.* This word came into Latin as *cubicus.* The good English adjective *cubic* is formed from the root of this Greek (and then Latin) adjective. Someone who knew some but not enough Latin then superimposed the Latin adjectival suffix *-alis* upon the stem of what was already an adjective to produce the low word *cubical.* These two words have both survived in common use, which rarely happens, since the bad word ending in *-ical* usually drove out the correct word ending in *-ic.*

cum The Latin preposition *cum* means *with.* It forms compounds according to the following rule:

> **Cum** appears as **con** before **t, d, c, q, g, s, f,** and **v**; as **com** before **p, b,** and **m.** Before **l** the unassimilated form is preferable except in **col-ligō** and its compounds, e.g. **con-locō, con-loquium, con-lapsus,** etc. But before **r** the assimilated form is preferable, as **cor-rumpō, cor-ripiō,** etc. Before vowels, **h,** and **gn** the form is **co,** as **co-alēscō, co-haereo, co-gnosco** (from **gnōscō,** the older form of **nōscō**). Before **n** the form is **cō,** as in

cō-nīveō, cō-nectō. **Comb-ūrō** is probably formed after the analogy of **amb-ūrō**. Before consonantal **i** the proper form is **con**, as **con-iungō, con-iūrō**, etc.; so **con-iciō** from **con-ieciō**, but also co-iciō (**30**, 1), like **co-alēscō**. (Hale and Buck, §51.6, p. 25)

cumulant The Latin noun *cumulus* means *a heap*. It is related to the Greek noun κῦμα, *swelling*, and the verbs κυέω and κύω, *to bear in the womb*. From the noun was formed the verb *cumulo, cumulare, cumulavi, cumulatus* with the meaning *to heap up*. *Cumulant* comes from the stem of the present participle *cumulans, cumulantis* of *cumulo*.

cumulative See the preceding entry. The Latin adjective *cumulativus* was formed by adding the suffix *-ivus* to the stem of the fourth principal part of the verb *cumulo*, a process that marks a word as likely medieval or modern.

curriculum Mathematicians often have to write a *curriculum vitae*. The plural is *curricula vitae* if the reference is to the activities of one person, *curricula vitarum* if one is referring to the activities of more than one person. Expressions such as *curriculum vita, curriculum vitaes* or *vitas*, or *curriculums vitae* are comical.

cursive The Latin verb *curro, currere, cucurri, cursus* means *to run*, and the addition of the verbal adjectival suffix *-ivus* to the stem of the fourth principal part produces the late Latin adjective *cursivus* with the meaning *having the quality or tendency of running*. The English adjective is used today solely as descriptive of a form of handwriting opposed to printing. This form of handwriting is being phased out of grade school instruction for two reasons: 1) It is imagined that the availability of computers makes cursive handwriting superfluous, and 2) no examinations for competence in cursive handwriting are required by the government, and therefore it is a luxury in which third-grade teachers cannot afford to indulge. The result will be the prolongation of mathematics lectures because the instructors will be printing everything, and printing takes more time than cursive script. The consequence for society will be the loss forever of a thing of beauty.

curtate This adjective was formed from the fourth principal part of the Latin verb *curto, curtare, curtavi, curtatus*, which means *to shorten*; the verb is derived from the adjective *curtus, short*, which is the same as the Greek κυρτός, *bent*.

curvature The Latin noun *curvatura* means *arching*. It is derived from the future active participle *curvaturus, -a, -um* of the verb *curvo, curvare, curvavi, curvatus, to bend*. If a plane curve has parametric equations $x = x(t)$ and $y = y(t)$, then the absolute curvature at a point is given by

$$\kappa = |x'y'' - y'x''| / [(x')^2 + (y')^2]^{3/2},$$

where the derivatives are with respect to the parameter t. If the curve has polar equation $r = r(\theta)$, then the absolute curvature is given by

$$\kappa = |r^2 - rr' + 2(r')^2| / [r^2 + (r')^2]^{3/2},$$

where the derivatives are with respect to θ.

curve The Latin adjective *curvus, bent*, is related to the Greek κυρτός with the same meaning. The noun *linea* of the phrase *linea curva, curved line*, was eventually dropped, and the remaining adjective soon took on the force of a noun. The British scientist Karl Pearson (1857–1938) pointed out that many of the most important probability density functions of mathematical statistics are solutions of the differential equation $y'/y = (d - x)/(a + bx + cx^2)$, $y > 0$, where a, b, c, and d are parameters. The solutions are called *Pearson curves*.

curvilinear This Latin-based word is correctly composed of *curvus* and *linearis*. See the entries **curve** and **linear**.

cusp The Latin noun *cuspis, cuspidis* means *the point*, especially *of a spear*. It is used for a point on a plane curve where the left and right derivatives both exist but are not equal.

CW complex The use of letters to name mathematical objects is to be deplored. If the letters ever had a meaning, they are forgotten in the next generation.

cycle The Greek noun κύκλος means *circle*. Its transliteration into Latin is *cyclus*, and that is the origin of our *cycle*.

cyclic This is the stem of the Greek adjective κυκλικός, *pertaining to a circle*.

cycloid The Greek word is κυκλοειδής, *shaped like a circle*, because of this curve's superficial resemblance to a circular arc. The Greek word for *circle* is κύκλος; the word εἶδος means *shape*. Since a wheel (τρόχος) has the shape of a circle, an old synonym for cycloid was *trochoid* from τροχοειδής, *shaped like a wheel*. It is the curve traced out by a fixed point on a circle that rolls without slipping on a straight line. The word *trochoid* now has the specialized meaning of the curve traced out by any fixed point connected to the center of the rolling circle. If the fixed point is within the circle, the trochoid is called *curtate* (shortened); if it is outside the circle, it is called *prolate* (extended). In both cases, the Latin adjectives are in uneasy alliance with the Greek noun. The curve was called *the Helen of geometers* because of the controversies aroused by the competition amongst mathematicians to discover its properties. Descartes (1596–1650) calculated the equation of the tangent line to the cycloid at an arbitrary point on it. Roberval (1602–1675) and Torricelli (1608–1647) proved that if the radius of the rolling circle is r, and one studies the arc with parametric equations

$$x = r(\theta - sin\ \theta) \text{ and}$$

$$y = r(1 - cos\ \theta), 0 \leq \theta \leq 2\pi,$$

the area of the plane region between the arc and the base is $3\pi r^2$. Wren (1632–1723) showed that the length of the arc is $8r$ and that the centroid of the arc is at $(\pi r, 4r/3)$. Pascal demonstrated that the centroid of the plane region is at $(\pi r, 5r/6)$ and that the volume of the solid produced by revolving the region about its base is $5\pi^2 r^3$. Fermat (1601–1665) showed that the surface area swept out by revolving the arc around the base is $64\pi r^2/3$. In Euler's Latin, the cycloid is *cyclois, cycloidis*.

cyclometric This is the learned name for the inverse trigonometric functions. It is an adjective and means *the circle-measuring*. It is the modern combination of the Greek noun κύκλος, *circle*, and adjective μετρικός, *pertaining to measure*, from μέτρον, *measure, standard*.

cyclotomic This adjective was used by Sylvester in the nineteenth century and is formed by adding the stem *-ic* of the Greek adjectival suffix -ικός to the stem of the noun *cyclotomy*. See the following entry.

cyclotomy This nineteenth-century noun is a combination of the two Greek words κύκλος, *circle*, and τομή, *a cutting*, and is meant to name the science of dividing a circle into a given number of equal parts.

cylinder The Romans transliterated the Greek noun κύλινδρος into their language as *cylindrus*, and the English word comes directly from the stem *cylindr-*.

cylindrical The Greeks added the adjectival suffix -ικός to the stem of κύλινδρος to form the adjective κυλινδρικός, "*cylindric.*" Modern authors (at least as early as the seventeenth century), unmindful that *cylindric* was already an adjective, then superimposed the Latin adjectival ending *-alis* on the stem of the Greek adjective to produce the low adjective *cylindricalis*, of which *cylindrical* is the stem. The real Latin word for this idea is *cylindratus, -a, -um*.

D

data This is the nominative plural of the Latin word *datum, something given*. The word therefore means *some things that are given*. It is a plural, and it must therefore be used with a plural verb. One says, "The data are...," not "The data is..." The hyphenated absurdities *data-centric* (*sc.* discipline), *data-enabled* (*sc.* sciences), and *data-intensive* (*sc.* research) appeared in an encyclical letter by Sastry Pantula, director of the Division of Mathematical Sciences of the National Science Foundation, quoted by George Andrews on page 933 of the August 2012 issue of the *Notices of the American Mathematical Society*. See the entry **cant.**

de- The Latin suffix *de-* is derived from the preposition *de* with the meaning *down from, following from, after*.

decrease The Latin verb *decresco, decrescere, decrevi, decretus* means *to grow* (*cresco*) *smaller* (*de-*).

decagon The Greek noun δεκάγωνον was the name of the equilateral polygon with *ten* (δέκα) *sides* (γωνία = *side*).

decameter This is a unit of the French revolutionary system of measurement and is equivalent to ten meters. The formation of this word is in accordance with knowledge. It is the combination of the numeral δέκα, *ten*, and μέτρον, *measure*. See the article **meter.**

decidable The Latin verb *decido, decidere, decidi, decisus* means *to cut* (*caedo*) *down or off* (*de-*), *to cut short, settle, arrange.* The suffix *-able* is the corruption of the Latin *-abilis.* A *decidable proposition* is one that can be proven true or false. See the entry **-able.**

decimal The Latin noun *decem* means *ten*; from this noun comes the adjective *decimus*, which means *tenth*. The medieval Latin word

102

decimalis meaning *pertaining to tenths or tithes* was formed by adding the adjectival suffix *-alis* to the noun *decem*.

decline The Latin first-conjugation verb *declino, declinare, declinavi, declinatus* means *to turn aside*. It is the compound of the preposition *de-* and the obsolete verb *clino, clinare*. The Latin verb is related to the Greek verb κλίνω, *to make slope or slant*.

decomposable This is a French word taken over into English. The late Latin verb *pauso, pausare, to stop*, became the French *poser*. The compounds *composer* and *décomposer* were then formed, and from the latter proceeded the adjective *décomposable*. This adjective is meant to describe that which has been put together and yet is capable of being taken apart. According to Weekley, *pauso* and its compounds managed in many cases to replace *pono* and its compounds. Had this word come directly from *compono*, it would have been *decomponable*.

decomposition The Latin verb *compono* means *to put (pono) together (cum)*. *Decompono* means *to take apart what has been put together*. The fourth principal part of the latter is *decompositus, taken apart*. From this adjective was formed the noun *decompositio, a taking apart*.

decreasing The Latin verb *decresco, decrescere, decrevi, decretus* means *to grow (cresco) down (de-), to become smaller*. The English word *decrease* arose from this Latin verb through the mediation of the Old French stem *descreiss-* of *descreistre*. The letter *a* started to appear in the spelling in the sixteenth century.

deductive The Latin verb *deduco, deducere, deduxi, deductus* means *to lead (duco) out (de-)*. The adjectival suffix *-ivus* was appended to the stem of the fourth principal part to produce *deductivus*, pertaining to consecutive thought, from whence came the English adjective.

deferent This is the name of a circle in the geocentric theory; it is the circle upon which the epicycle rotates. See **epicycle**. It is derived from the present participle *deferens, deferentis* of the Latin verb *defero*, *deferre, detuli, delatus,* which means *to bring or carry (fero) down (de-) or away, to convey.*

deficient The Latin verb *deficio, deficere, defeci, defectus* means *to do less than one could.* The present participle is *deficiens, deficientis,* and from this participle comes the English adjective *deficient.* A *deficient number* is a number the sum of whose proper divisors is less than the number itself. Such numbers were first discussed by Nicomachus (*circa* 80–120) in his *Introduction to Arithmetic,* where they are called ἀριθμοὶ ἐλλιπεῖς.

definite The Latin verb *finio, finire, finivi, finitus* means *to bound,* and the compound *definio, definire, definivi, definitus* means *to set bounds to, to define.* It is from the fourth principal part of this verb that the word *definite* proceeds. The force of the *de-* is to emphasize the thoroughness of the action, *to bound utterly.* The plural noun *fines* means *boundaries.*

deformation The Latin verb *deformo, deformare, deformavi, deformatus* means *to put something out of (de-) its proper shape (forma).* From the fourth principal part was produced the noun *deformatio, deformationis,* from whose stem proceeded the French and English nouns.

degenerate The Latin adjective *degener, degeneris* means *fallen away from (de-) one's origin (genus), unworthy of one's race.* From this adjective was formed the verb *degenero, degenerare, degeneravi, degeneratus* with the meaning *to become unlike one's kind.* The English adjective is derived from the fourth principal part of the Latin verb.

degree It is supposed by Weekley *sub voce* that the Latin noun *gradus, step,* was at some point augmented by the addition of the prefix *de-* to *degradus,* whence the English word arose. The form *degradus* is unrecorded in literature.

del This is the clumsy abbreviation of the name of the Greek letter *delta.* Such abbreviations are poor style and are to be condemned. They are like nicknames, unsuitable for anything above the level of colloquial speech. The *del* is supposed to mean the upside-down *delta* ∇ used by Hamilton to indicate the gradient of a function of three variables. Its use is now established by immemorial custom and may be tolerated, as may similarly traditional misoriented letters such as $\forall$ and $\exists$. However, such notation should not be multiplied by modern authors; *quod licet Iovi non licet bovi*. The symbol ∇ is by some poorly advised people referred to as *atled*, the word *delta* spelled backwards, and by more affected individuals as *nabla*, the supposed name of a similarly shaped instrument of the ancient Israelites. However, Brown, Driver, and Briggs are unsure whether the instrument in question is a lute, guitar, or harp, and therefore hypotheses about its shape must remain dark and doubtful. Furthermore, the word is נֶבֶל, *nēbĕl*, not *nabla*. Thus, both the name *atled* and the name *nabla* are the offspring of folly and ignorance.

delta *Delta* is the fourth letter of the Greek alphabet, used in mathematics in the small case δ for an arbitrary small positive real number and in the capital Δ for the discriminant of the equation of a conic section. The δ-function is the extended real valued function defined on the real numbers by $\delta(x) = 0$ if $x \neq 0$, $\delta(0) = \infty$; it is assumed to satisfy certain other conventions, which may be found in *Problems in Probability Theory, Mathematical Statistics, and Theory of Random Functions* by A. A. Shveshnikov, Dover Publications, Inc., New York, 1978, page 49.

deltoid This is the name of a curve that looks like Δ, a capital delta. The Greek word is δελτοειδής, a compound of δέλτα, the letter Δ, and εἶδος, which means *shape*.

denominator The Latin verb *denomino, denominare, denominavi, denominatus* means *to give a name (nomen) to someone*. The addition of the nominal suffix *-or* to the stem of its fourth principal part produces the noun *denominator, one who gives a name*. The use of this noun for the second part of a fraction is at least as old as the sixteenth century. The prefix *de-* in this case intensifies the idea of the naming, to give a name so as to leave no doubt what something is.

105

dense The Latin adjective *densus* means *thick, close, compact.*

density From the Latin adjective *densus* meaning *thick, close* was formed the noun *densitas* by the addition of the suffix *-itas* to the stem of the adjective. This noun entered French as *densité* and from there came into English as *density.*

denumerable This adjective came into English via French from the Latin *denumerabilis, capable of (-abilis) being numbered (denumero = to number,* from *numerus, number).* The word describes a set whose elements can be put into one-to-one correspondence with a subset of the positive integers.

dependence See the entries **dependent** and **independent**. The "word" *dependentia,* whether of individuals or of nations, never existed in Latin. The Romans used the word *clientela* for this idea.

dependent The Latin verb *dependeo, dependere* means *to hang (pendeo) down (de-), to hang from.* It has the present participle *dependens, dependentis,* from whose stem is derived the English adjective.

derivative The Latin verb *derivo, derivare, derivavi, derivatus* means *to turn away (de-) into another channel (rivus), to divert.* The adjectival suffix *-ivus* was added to the stem of the fourth principal part of the verb to produce the adjective *derivativus* (pertaining to what has been diverted into another channel, to what passes from one thing to another and therefore derivative), a word used by Priscian *circa* A.D. 500, whence came the English and French adjectives. Lagrange (1736–1813) spoke of derivatives in the modern sense in his *Théorie des functions analytiques.*

design The Latin verb *designo, designare, designavi, designatus* means *to mark (signo) out (de-).* The noun *signum* means *a mark.*

determinant The Latin verb *determino, determinare, determinavi, determinatus* means *to fix the limits of something,* from *terminus, limit.* Its present participle is *determinans, determinantis,* and the English word in question is the stem of this participle. The prefix *de-* in this case intensifies the idea of the fixing of the boundaries, to give the limits so as to leave no doubt where the boundaries are.

deviation [standard] The Latin noun *via* means *the way*. The late Latin verb *devio, deviare, deviavi, deviatus* means *to depart from (de-) the way*. From the fourth principal part of this verb was produced the noun *deviatio, deviationis*, which means *a departure from the way*.

diagonal This is a low word, the Latin adjectival suffix *-alis* having been appended to the Greek adjective διαγώνιος, *leading across*, derived from the verb διάγω, *to lead across*. This word was of course not used by Euclid, who used the term *diameter* for what we call a diagonal of a square, rectangle, or parallelogram.

diagonizable The suffix *-able* of Latin origin has been added to the stem of a Greek verb διαγονίζω to produce an adjective meaning *capable of being put into diagonal form*.

diagram The Greek noun διάγραμμα, διαγράμματος is *that which is marked out by lines*. The technical term used by Euclid for a geometrical figure, however, was σχῆμα.

diameter The Greek noun ἡ διάμετρος is the technical term used by Euclid for *a chord through the center of a circle*. The related noun τὸ διάμετρον is *the portion measured* (μέτρον = *measure or rule*) out (διά). The preposition διά has the sense of distribution when appended to another word. Boëthius merely transliterated the Greek word when he prepared his translation of Euclid's *Elements*. Transliterations are frequently used out of respect for established technical terms, or to express ideas that do not exist in the new language. They are also employed when the translator is intellectually limited and does not understand what he is translating. A spectacular example of a succession of transliterated words appears in Adelard's Latin translation of Proposition 2 of the spurious Book XV of al-Hajjaj's Arabic edition of Euclid:

Itaque factum est in almugecum ABGD quattuor alkaidarum almugecum habens octo alkaidas triangulorum equalium laterum.

And so there has been produced in the almugecum ABGD of four alkaidas an almugecum with eight alkaidas that are equilateral triangles.

dice This is the plural of the singular *die*; these words are the corruption of the Latin *datum*, *a given*, from the verb *do, dare, dedi, datus, to give*.

Dido In Books II–IV of his *Aeneid*, Vergil tells the story of Dido, queen of Carthage, who fell in love with the hero Aeneas. The episode contains a famous mathematical problem.

> Devenere locos, ubi nunc ingentia cernes
> Moenia surgentemque novae Carthaginis arcem,
> Mercatique solum, facti de nomine Byrsam,
> taurino quantum possent circumdare tergo.
> (Aeneid, Book I, 365–368)

> At last they landed, where from far your Eyes
> May view the Turrets of new Carthage rise:
> There bought a space of Ground, which Byrsa call'd
> From the Bulls hide, they first inclos'd, and wall'd.
> (Dryden's translation, Book I, 505–508)

The problem is: Given a string of length πr whose endpoints are attached to a line of length $2r$, what arc should the string describe so as to maximize the area of the enclosed figure? The answer is that the string should describe a semicircle. This was an early example of an *isoperimetric* problem.

diffeomorphism This bad word was meant to be a combination of *differentiable* and *homeomorphism* and to indicate a homeomorphism that was differentiable. It is the same sort of word as *mathlete, q.v.*

difference The Latin verb *differo, differre, distuli, dilatus* means *to carry (fero) in different directions (dis-), to take apart*. From its present participle *differens, differentis* is derived the noun *differentia* with the meaning *difference, distinction.*

differentiable See the entry **differentiate**.

differential See the entry **differentiate**. A *differential equation* is an equation that contains a derivative. The differential equation

$$M(x,y) + N(x,y)\,y' = 0$$

is often written in the form

$$M(x,y)dx + N(x,y)\,dy = 0.$$

The expression $M(x,y)dx + N(x,y)\,dy$ is called a *differential form.*

differentiate The modern Latin word *differentio, differentiare, differentiavi, differentiatus* was invented by Leibnitz in 1677 to indicate what we would call the limit of the ratio of differences. From it were derived in the usual manner the associated noun *differentiatio, differentiationis* and the adjectives *differentiabilis* and *differentialis.*

differentiation See the entry **differentiate**.

digit The Latin noun *digitus* means *finger* or *toe*. Since the decimal system of numbers arose on account of our having ten fingers, the word *digitus* acquired its mathematical sense of *a symbol for a number.*

digital The addition of the suffix *-alis* to the stem of the noun *digitus* produced the adjective *digitalis* with the meaning *pertaining to the finger.*

digraph The English prefix *di-* may be the Latin prefix *di-* shortened from *dis-* or the Greek prefix δι-, shortened from δίς, *twice*. In the latter case it corresponds to the Latin *bi-*, which is abbreviated from *bis*, which means *twice*. It is the Greek prefix that we find in this word, along with the noun γραφή, *writing*, from the verb γράφω, *to write or draw.*

dihedral This strange word consists of a Latin suffix *-alis* pasted on a Greek noun δίεδρα, *a seat with room for two.*

dilation The Latin noun *dilatio, dilationis* is derived from the fourth principal part of the verb *differo, differre, distuli, dilatus*, which means *to carry (fero) in different directions (dis-), to take apart.* The noun means *a putting off, a delaying, a postponing.* From the fourth principal part was derived a second frequentative verb *dilato, dilatare, dilatavi, dilatatus* with the meaning *to spread out, to make wide (latus = wide).* The mathematical term *dilation* is actually derived from this second verb, but the derivation was not done correctly. The word should have been *dilatation.*

dimension The Latin verb *dimetior, dimetiri, dimensus* means *to measure out*; it is the combination of the inseparable prefix *dis-* and the verb *mentior, mentiri, mensus*, which means *to measure.* From the participle *dimensus* was formed the noun *dimensio, dimensionis* with the meaning *a measuring.*

Diophantine The name of the Greek mathematician Διοφάντης (third century A.D.) became *Diophantus* in Latin, and the addition of the adjectival suffix *-inus* produced the adjective *Diophantinus*, whence came the English *Diophantine*. In a similar manner we get *Philippine* from the Greek name Φίλιππος.

direct The Latin verb *dirigo, dirigere, direxi, directus* means *to arrange, make straight.* It is the combination of the inseparable prefix *dis-* and the verb *rego, regere, rexi, rectus*, which means *to make go straight, to guide, direct, rule.* The participle *directus* became a common word for *straight.*

directional The Latin noun *directio, directionis* was derived from the fourth principal part of the verb *dirigo*; see the entry above. The adjectival suffix *-alis* was added to the stem of the noun to produce the adjective *directionalis*, whose stem is the English adjective.

director [circle] The nominal suffix of agent *-or* was added to the stem of the fourth principal part *directus* of the verb *dirigo* to produce the noun *director, directoris* with the meaning *the one who arranges or straightens out.* The noun is masculine because it modifies *circulus,* which is masculine. If an ellipse has semi-major axis *a* and semi-minor axis *b,* then any two of its tangents that are at right angles intersect on a circle, called the *director circle,* whose center is the center of the ellipse and whose radius is $(a^2 + b^2)^{1/2}$.

directrix This is the feminine form in Latin of the masculine noun *director.* It is a noun of agent formed from the fourth principal part of the verb *dirĭgo, dirigere, direxi, directus,* which means *to arrange,* which is the combination of the inseparable prefix *dis-* (*apart*) and the verb *rego* (*to rule*). It is used of the given line in the *focus-directrix* definition of the conic sections. It is feminine because it modifies the noun *linea* (*line*), understood, which is feminine.

dis- The Latin inseparable prefix *dis-, di-, dir-* conveys the idea of *separately, apart, in different directions.* It indicates disassociation from whatever follows. It is related to *bis, twice,* and to δύο, *two.* It is sometimes modified to *de-* in English; for example, *disembark* became *debark,* and *disarrange* became *derange.*

disc The Greek noun δίσκος, *the round plate that their athletes threw,* was transliterated into Latin as *discus.* One writes *disk* or *disc* in accordance with whether one transliterates from the Greek in the manner of George Grote or from the Latin. The second alternative is the older and more correct. The Latin letter *c* always transliterates the Greek *kappa* and never the Greek *sigma,* which went into Latin as *s.* There is therefore no confusion in adhering to the traditional procedure. *Disc* is thus preferable to *disk.* Grote discussed the transliteration of the Greek letter **K** in the prefatory statement *Names of Gods, Goddesses, and Heroes* on pages xxiii–xxiv of volume I of the fourth edition of his *History of Greece* (John Murray, London, 1854):

> A few words are here necessary respecting the orthography of Greek names adopted in the above table and generally

throughout this history. I have approximated as nearly as I dared to the Greek letters in preference to the Latin; and on this point I venture upon an innovation which I should have little doubt of vindicating before the reason of any candid English student. For the ordinary practice of substituting, in a Greek name, the English C in place of the Greek **K** is indeed so obviously incorrect, that it admits of no rational justification. Our own **K** precisely and in every point coincides with the Greek **K**: we have thus the means of reproducing to the eye as well as to the ear, yet we gratuitously take the wrong letter in preference to the right. And the precedent of the Latin is here against us rather than in our favour, for their C really coincided in sound with the Greek **K**, whereas our C entirely departs from it, and becomes an S, before *e*, *i*, *æ*, *œ*, and *y*. Though our C has so far deviated in sound from the Latin C, yet there is some warrant to continue to use it in writing Latin names—because we thus reproduce the name to the eye, though not to the ear. But this is not the case when we employ our C to designate the Greek **K**, for we depart here not less from the visible than from the audible original; while we mar the unrivalled euphony of the Greek language by that multiplied sibilation which constitutes the least inviting feature in our own. Among German philologists the K is now universally employed in writing Greek names, and I have adopted it pretty largely in this work, making exceptions for such names as the English reader has been so accustomed to hear with the C, that they may be considered to be almost Anglicised. I have farther marked the long *e* and the long *o* (η, ω) by a circumflex (*Hêrê*) when they occur in the last syllable or in the penultimate of a name.

disconnected This adjective is composed of the prefix *dis-* and the adjective *connected*. See both those entries. The formation of verbs by adding the prefix *dis-* to other verbs of Latin origin is standard practice and correct. The participles of the new verbs are then commonly used as adjectives.

disconnection This modern word (it has been around since at least the eighteenth century) was formed as if there were a Latin noun *disconexio, disconexionis*. It is unobjectionable. The doubling of the *n* occurred during the transition from classical to medieval Latin.

discontinuity This noun is the composition of the prefix *dis-* and the noun *continuity*. See both those entries.

discrete The Latin verb *discerno, discernere, discrevi, discretus* means *to sever, separate, figure out (cerno) by taking apart (dis-)*. The adjective *discrete* is formed from the fourth principal part. The word *discreet*, which now is reserved for the description of prudent behavior, actually began as an alternative English spelling of *discrete*. The mathematical term is the etymologically correct spelling *discrete*. The vowels and diphthongs *e, ee, ea* are all pronounced the same way in the words *discrete* (the first *e*), *discreet*, *concede* (the first *e*), *proceed*, *east*, and *eat*.

discriminant The Latin noun *discrimen, discriminis* means *that which divides*. It is the compound of the prefix *dis-* and *crimen*, the latter a noun derived from the verb *cerno, cernere, to separate, to distinguish*. From *discrimen* was derived the verb *discrimino, discriminare, discriminavi, discriminatus* with the meaning *to separate, sunder, divide*. The stem of the present participle *discriminans, discriminantis* is the English word *discriminant*. The discriminant of a conic section with equation $Ax^2 + Bxy + Cy^2 + Dx + Ey + F = 0$ is $B^2 - 4AC$. The discriminant is invariant under translation and rotation of axes. If it is positive, the conic is a hyperbola or two intersecting lines; if it is zero, the conic is a parabola or two parallel lines; if it is negative, the conic is an ellipse, a circle, two coincident lines, a point, or an imaginary ellipse.

disjoint The Latin verb *disiungo, disiungere, disiunxi, disiunctus* means *to take apart that which was joined*. There was in French the progression *disiunctus, disjunctus, disjunct, disjoinct, disjoint*.

disk See the entry **disc** above.

dispersion This is another name for what is more commonly called the *variance*. The Latin verb *dispergo, dispergere, dispersi, dispersus* means *to scatter, to sprinkle (spergo) all over the place (dis-)*. The noun is the stem of the Latin *dispersio, dispersionis*, formed from the fourth principal part.

dissimilar The Latin adjective *dissimilis* means *unlike* and is formed from the prefix *dis-* and the adjective *similis*. See the entries **dis-** and **similar**. The adjective *similis* is derived from the adverb *simul*, which means *together*. From the adjective was produced the verb *simulo, simulare, simulavi, simulatus* with the meaning *to make like*. The existence of the French adjective *similaire* indicates that there was in late use the Latin adjective *simularis*.

dissipative The Latin verb *dissipo, dissipare, dissipavi, dissipatus* means *to scatter, disperse, throw about*; it is the compound of the prefix *dis-* and the rare verb *sipo, sipare, sipavi, sipatus, to scatter*. From it were derived the noun *dissipatio, dissipationis*, meaning *scattering*, and the adjective *dissipabilis*, which means *that which can be scattered*. The form *dissipativus* never existed in Latin, but it is a good and natural product of the formation of Latin verbal adjectives. See the entry **-ive**.

distance The Latin verb *disto, distare* means *to stand (sto) apart (dis-)*. Its present participle is *distans, distantis*, and from it is derived the noun *distantia*, which came into French and then English as *distance*.

distinguishable The Latin verb *stinguo, stinguere* means *to extinguish, to annihilate*, and is related to the Greek verb στίζω, *to prick, tattoo, brand*. The addition of the prefix *dis-* produced the compound verb *distinguo, distinguere, distinxi, distinctus* with the meaning *to mark off, separate*. The word *distinguishable* should have been *distinguable*; the *ish* is a mistake, as in the cases of *admonish*, *astonish*, and *exstinguish*, which examples are adduced by Weekley. It is the metamorphosis of the French *-iss-* added in the formation of the present participle of verbs ending in *-ir*, as *agissant* from *agir*. The verb *distinguer* was evidently at some time imagined by the *misera plebs* to be *distinguir*.

distribution The Latin noun *distributio, distributionis* is derived from the fourth principal part of the verb *distribuo, distribuere, distribui, distributus*, which means *to divide (tribuo) among (dis-)*.

distributive By adding the suffix *-ivus* to the stem of the fourth principal part of the verb *distribuo* (see previous entry), there was produced the adjective *distributivus* with the meaning *tending to distribute*.

divergence The Latin verb *vergo, vergere, versi* means *to turn*. By the addition of the prefix *di-* (from *dis-*) there was produced the verb *divergo* with the meaning *to turn away from*. The English noun is derived from the present participle *divergens, divergentis*.

divergent This adjective is derived from the present participle *divergens, divergentis* of the Latin verb *divergo*, which means *to turn away from*.

divide The English verb comes from the first principal part of the Latin verb *divido, dividere, divisi, divisus*, which means *to separate into parts*.

dividend This noun is derived from the gerund *dividendus* of the Latin verb *divido, dividere*. It means *that which is to be divided*.

divisibility The Latin noun *divisibilitas, the ability to be divided*, came into French with the *-as* transformed into *é*, and the English eventually replaced the *é* by *y*, as usual.

division The Latin noun *divisio, divisionis* means *the act of breaking up into parts*. It came into French and English as *division*.

divisor This Latin noun means *someone who divides*. It is formed by adding the nominal suffix *-or* to the stem of the fourth principal part of the verb *divido, dividere, divisi, divisus*, which means *to separate into parts*.

dodecagon This is the Latin transliteration of the Greek noun δωδεκάγωνον from δώδεκα, *twelve*, and γωνία, *angle*. The latter noun is akin to γόνυ, *knee*, for the angle has the shape of the bent knee.

dodecahedron This is the Latin transliteration of the Greek word δωδεκάεδρον for the Platonic solid with twelve faces. *Dodeca* is the

Greek δώδεκα, *twelve*, and *hedron* is the Greek ἕδρον, *seat*. Grote, in the preface of his *History of Greece*, argued for transliterating the Greek *kappa* by *k* instead of *c*, but this should not be done in the case of words that were adopted into the Latin language before English came to be.

domain This is the transformation of the Latin noun *dominium*, the property of the *dominus* or lord. *Dominium* became *domaine* in French, and then the English dropped the unpronounced final *e*. This word is now an entry in the dictionary of internet cant: "The system cannot log you on because the domain is unavailable" means that the computer will not work because you made a mistake in typing in the email address.

dominant The Latin verb *domino, dominare, dominavi, dominatus* means *to lord it over*. Its present participle is *dominans, dominantis*, whence comes the English adjective *dominant*.

double This is the metamorphosis of the Latin adjective *duplus*, related to the Greek διπλόος, contracted to διπλοῦς, *twofold*. The *plus* part of the word is from the Latin *plenus*, related to the Greek πλέος, *full*.

droid Like all computer and technology cant, this is a concoction of modern enthusiasts with no education in literature, for the syllable *droid* suggests nothing but illiteracy to the learned. There is nothing *soft* about *software*, nothing *hard* about *hardware*, nothing *up* about *upload*, and nothing *down* about *download*. The meanings assigned by uncultured people must simply be memorized.

dual The Latin word for *two* is *duo*, the same as the Greek δύω. From this noun, upon the addition of the adjectival suffix *-alis* to the stem, is formed the adjective *dualis, pertaining to two*. The English adjective is formed by removing the nominative case ending *-is*.

duality The Latin noun *dualitas* with the meaning *twoness* is formed from the adjective *dualis* by adding the nominal suffix *-itas* to the stem. Nouns like this came into French with the *as* replaced by *é*, producing in this case *dualité*. After these words crossed the Channel, the *é* eventually became *y*.

duodecimal From the Latin noun *duodecem, twelve,* comes the adjective *duodecimalis* with the meaning *pertaining to twelve*, whence proceeds the English *duodecimal.*

duplication [of the cube] The Latin noun *duplicatio, duplicationis* means *doubling.* It comes from the verb *duplico, duplicare,* which is itself derived from *duo, two,* and *plico, plicare, to fold* (related to the Greek verb πλέκω with the same meaning). Thus, the word meant originally *to fold so as to make two.*

dyadic The Greek noun δυάς, δυάδος means *the number two.* Adding the adjectival suffix -ικός to the stem produces the adjective δυαδικός with the meaning *pertaining to two.* The transliteration *dyadik* became *diadick* and then *dyadic.*

dynamical The Greek noun δύναμις means *power.* The corresponding Greek adjective is δυναμικός, *pertaining to power.* The correct English adjective is therefore *dynamic.* To superimpose the vestige -*al* of the Latin adjectival ending -*alis* upon the stem of a Greek adjective is often the product of ignorance and produces a low word. In other cases, the addition of the Latin suffix to the Greek adjective is due to the fact that a different meaning is intended from that of the Greek adjective; thus, *dynamic* was an established word, so one spoke of *dynamical systems* rather than *dynamic systems* to avoid confusion.

E

e The letter *e* is the symbol for the Eulerian constant, 2.718281828 to nine decimal places. The mnemonic device, of use only to those competent in American history, is that 1828 was the year when Andrew Jackson was elected president of the United States.

eccentricity This is the measure of how different a conic section is from a circle. The eccentricity of the Colosseum is .5588; that of St. Peter's Square, is .708. The Greek adjective ἔκκεντρος means *off center*.

echelon This is the French corruption of the Latin *scala, staircase*. *Échelon* is French for *the rung of a ladder*.

ecliptic The Greek verb ἐκλείπω means *to leave* (λείπω) *off* (ἐκ), *to come short*. The associated noun is ἐκλείψις, which means *a coming short, a falling off*. From this noun was formed the associated adjective ἐκλειπτικός with the meaning *pertaining to a falling off*. The ecliptic is the apparent path of the sun through the starry heavens, so-called because an eclipse is only possible when the moon is near this path.

efficient [estimator] The Latin verb *efficio, efficere, effeci, effectus* means *to do, produce, make*. Its present participle *efficiens, efficientis* means *bringing about*. Our adjective is just the stem of this participle. The verb *efficio* is the compound of the preposition *ex* meaning *from, out of*, and the verb *facio, facere, feci, factus, to do*.

elastic This is the stem of the Greek adjective ἐλαστικός, *propulsive*, formed from the addition of the adjectival suffix -ικός to the stem of the noun ἐλαστήρ, *a driver*, from the verb ἐλαύνω, *to propel, to drive forth*.

election This is the Latin noun *electio, electionis*. The noun is formed from the fourth principal part of the verb *eligo, eligere, elegi, electus*, which means *to choose* (*lego*) *out of* (*e-*), *to elect*. Election or ballot problems form an indispensable part of any course on probability. The solution of ballot problems suggested by episodes to be found in history books is a profitable diversion. They indicate how the highest authorites in other subjects may demonstrate a credulity impossible in a mathematician. The following stories are related by Dean Milman in volume 7, pages 121 and 538 of his *History of Latin Christianity* (Sheldon & Company, New York, 1861):

> In the play of votes, now become usual in the Conclave, all happened at once to throw away their suffrages on one for whom no single vote would have been deliberately given. To his own surprise, and to that of the College of Cardinals and of Cristendom, the White Abbot, the Cistercian, James Fournier found himself Pope.

> The contest lay between a Spaniard and a French prelate. Neither would make concessions. Both parties threw away their suffrages on one whom none of the College desired or expected to succeed: their concurrent votes fell by chance on the Cardinal of Siena.

A similar event is claimed almost to have happened by a less reliable source:

> Twenty-five cardinals enetered the conclave. The absence of the French element left practically only two contending parties—the young and the old. The former had secretly settled on Giovanni de' Medici; the second openly supported S. Giorgio, England's candidate...The Sacred College had been assembled almost a week before the first serious scrutiny took place. Many of the cardinals, wishing to temporize and conceal their real intentions, had voted for the man they considered least likely to have any supporters. As luck would have it, thirteen prelates had selected the same outsider, with the result that they all but elected Arborense, the most worthless nonentity present. This narrow shave gave the Sacred College such a shock that its members determined to come to some agreement which would put matters on a more satisfactory basis for both parties. (V. Pirie, *The Triple Crown*, G. P. Putnam's Sons, New York, 1936, p. 49)

For more on this topic, see Chapter 19 of Paul J. Nahin, *Digital Dice, Computational Solutions to Practical Probability Problems*, Princeton University Press, 2008.

element This is the stem of the Latin noun *elementum*. On the etymology of this word, Allen and Greenough write (§239), "So *elementum* is a development from L-M-N-a, *l-m-n's* (letters of the alphabet [or, as we would say, *a-b-c's*]), changed to *elementa* along with other nouns in *-men*." It was used by the Romans to translate the title of Euclid's book, Στοιχεῖα.

elementary This Latin original of this word is formed by the addition of the adjectival suffix *-arius* to the noun *elementum* once the case ending *-um* has been removed.

elimination The Latin noun *limen, liminis* means *the threshold, doorway*. From it and the preposition *e, out*, was formed the verb *elimino, eliminare, eliminavi, eliminatus* with the meaning *to carry out of doors*. From the fourth principal part came the noun *eliminatio, eliminationis, a carrying out of doors*.

ellipse The ellipse is the set of all points in the plane, the sum of whose distances from two fixed points is fixed. The constant sum of the distances is usually denoted by *2a*, and the distance between the two fixed points is usually denoted by *2c*. The parameter *a* must be greater than the parameter *c*, so we define a third parameter *b* by $b^2 = a^2 - c^2$. The ratio *2c/2a* is called the eccentricity *e* of the ellipse. We must have $0 \le e < 1$, with equality on the left in the case when the two fixed points are the same. The *latus rectum* is $2b^2/a$. The Greek verb ἐλλείπω means *to leave* (λείπω) *in* (ἐν), *to leave behind, to leave out, to come short*. The associated noun is ἔλλειψις, which means *a deficiency, a coming short, a falling off*. The origin of the name is as follows. Consider the ellipse whose equation is $[(x - a)^2/a^2] + y^2/b^2 = 1$. Let *P(x,y)* be a point on the ellipse not a vertex, and let *S* be a square of side $|y|$. Let R be a rectangle whose base is *x* and whose altitude is the length of the latus rectum of the ellipse. Then the area of *S* is less than [that is, falls short of] the area of R. The ellipse may also be defined by the focus-directrix definition: Let $1 > e > 0$, let ℓ be a fixed line (the *directrix*) and F a fixed point (the *focus*) a distance d (the *directral distance*) from ℓ, $d > 0$. Then the ellipse is the locus of all points P such that $FP/F\ell = e$. If F is the pole and ℓ is the line with equation $x = d$, then the polar equation of the ellipse is $r = ed/(1 + e \cos\theta)$. The directral distance is related to the other parameters by the formula $d = a(1 - e^2)/e$. Gian Lorenzo Bernini made use of an ellipse of eccentricity .708 for St. Peter's Square, the most beautiful public space in the world. If one stands at a focus, the four rows of columns on the side of that focus appear as one.

ellipsoid This word is a modern invention made on the analogy of the Greek words *rhomboid* and *trapezoid*, the result of adding the suffix *-oid* to the stem of the noun *ellipse*. The word *ellipse* is taken from the Latin *ellipsis*, the corruption of the Greek ἔλλειψις; εἶδος is Greek for *shape*. The word *ellipsoid* should therefore mean *something that looks like an ellipse*. However, since the ellipsoid, being solid, does not look like an ellipse, which is a plane figure, the word does not convey the meaning it was intended to. Observe that the real Greek words *rhomboid* and *trapezoid* referred to plane figures, not solids.

elliptic This is the transformation of the Greek adjective ἐλλειπτικός, *pertaining to the ellipse, q.v.*

elliptical This adjective is the result of superimposing the Latin adjectival suffix *-alis* on top of the Greek adjective ἐλλειπτικός.

emeritus The Latin verb *mereo, merere, merui meritus* means *to deserve*, and the compound verb *emereo, emerere, emerui, emeritus* means *to deserve thoroughly, to obtain by service, to deserve very well of someone*. The force of the preposition *e* added as a prefix is to emphasize the completeness of the act. The past participle *emeritus* means *a soldier who has served his time, a veteran*. A *professor emeritus* is therefore a *retired professor*. A retired female professor is also a *professor emeritus*, and the use of the term *professor emerita* is a blunder comical to those who know Latin.

empirical The Greek adjective ἐμπειρικός means *experienced, trained.* It is derived from the noun πεῖρα, meaning *trial, experiment,* to which the prefix ἐν- (*in-*) has been added. The correct adjective is *empiric,* but people ignorant of Greek superimposed the Latin suffix *-alis* on the stem of what was already an adjective to produce *empirical.*

endomorphism This is a very modern word. The Greek adverb and preposition ἔνδον means *within, at home.* The noun μορφισμός, *a shaping or forming,* is derived from the noun μορφή, *a form, shape.* An endomorphism is a function from a set into itself that preserves the natural structure. Thus, an endomorphism of a group into itself is an automorphism. An endomorphism of a vector space into itself is a linear transformation. If the set has no structure, then the endomorphism is merely a function from the set into itself.

energy The Greek preposition ἐν means *in,* and the noun ἔργον means *work.* From these two words is formed the compound adjective ἐνεργός, *working,* and then the noun ἐνέργεια, *an action, operation, energy.*

entire This adjective came into English from the Latin *integer,* which means *whole,* through the mediation of the French *entier.*

entropy This is the Greek noun ἐντροπή, *a turning inwards, an inversion,* derived from the verb ἐντρέπομαι, *to turn oneself towards, to pay heed to.* The word was adopted by Clausius (1822–1888) for his principle that heat cannot of itself pass from a colder to a hotter body; evidently such a passage was viewed as an inversion of the natural law. The notion of entropy was introduced into mathematics by Kolmogorov and Sinai from the information-theoretic entropy of Claude Shannon. The entropy of a partition $\xi = \{A_1, A_2,...,A_n\}$ of the unit interval is defined to be

$$- [\mu(A_1) ln\, \mu(A_1) + \mu(A_2)\, ln\, \mu(A_2) + \mu(A_3) ln\, \mu(A_3) + \cdots$$

$$+ \mu(A_n) ln\, \mu(A_n)];$$

it is a measure of the uncertainty as to which of the partition elements an arbitrarily chosen number $x \in [0,1]$ belongs. The greater the uncertainty, the greater the entropy See Chapter X of "On the Origin and History of Ergodic Theory" by A. Lo Bello, *Bollettino di Storia delle Scienze Matematiche*, vol. I, no. 1, 1983, pages 37–75.

enumerable The Latin verb *enumero, enumerare, enumeravi, enumeratus* means *to reckon, to count up*. It is composed of the preposition *e* and the denominative verb *numero, numerare, to number*. The prefix *e* emphasizes the thoroughness of the action of the following verb. The addition of the suffix *-abilis* to the stem of the first principal part produced the adjective *enumerabilis* with the meaning *capable of being reckoned up*.

enumerate The English verb is derived from the fourth principal part of the Latin verb *enumero*, discussed in the preceding entry.

enumeration The Latin noun *enumeratio, enumerationis* means *a counting up*. It is derived from the fourth principal part of the verb *enumero* by the addition of the nominal suffix *-io* to the stem. See the entry **enumerate** above.

enunciation The Latin verb *enuntio, enuntiare, enuntiavi, enuntiatus* means *to tell, divulge, disclose*. It is composed of the preposition *e* (*out*) and the verb *nuntio*, which means *to announce*. From its fourth principal part was derived the technical term *enuntiatio, enuntiationis* in rhetoric with the meaning *proposition*.

epicycle The word *epicycle* is the Greek ἐπίκυκλος, from the preposition ἐπί, which means *on*, and the noun κύκλος, which means *circle*. It is the circle whose center rotates around another circle. The locus of a point on it was the basic curve of the Ptolemaic system and indeed of astronomy itself until the time of Kepler. The ancients attempted to describe all the motions observed in the heavens by epicycles. The practice of explaining the heavenly trajectories in terms of these curves was adopted by Ptolemy of Pelusium (second century A.D.) in his exposition of the geocentric theory as well as by Copernicus (1473–1543) in his treatment of the heliocentric system. This practice was founded upon the metaphysical assumption that the circle is the perfect curve, most appropriate for use by the Almighty, and therefore all celestial motions must be combinations of motions on circles. This assumption was not put aside until the time of Johannes Kepler (1571–1630), whose ellipses had hitherto been disqualified from consideration because they were not constructible with straightedge and compass. Let the origin O ("the center of the world") be the center of a circle of radius r_1. This circle is called the *deferent* from the Latin *defero*, which means *to convey*. Let the point Q travel on the perimeter of the deferent at constant angular speed s_1. If we assume that at time 0 the point Q is at the point $N = (r_1, 0)$, then we have $\theta = s_1 t$, where θ is the angle NOQ. The requirement of constant speed, which is known as the *regularity requirement*, was another metaphysical requirement that the ancients felt obliged to make. Let Q itself be the center of a circle of radius r_2. This second circle is called the *epicycle*. Let P be a point on the epicycle, and suppose that it travels on the perimeter at constant angular speed s_2. If we assume that the point P was at $(r_1 + r_2, 0)$ at time $t = 0$, and if we draw QM parallel to the x-axis and put $\varphi = angle\ PQM$, then we have

$$\varphi = angle\ PQM = s_2 t = s_2\, \theta / s_1,$$

or, if we set k equal to the ratio s_2 / s_1 of the angular speeds, we get $\varphi = k\theta$. The locus of P under this compound motion of P around Q and Q around O is the epicyclic curve. If we drop a perpendicular from Q to the x-axis meeting it at J, and another perpendicular from P to QM meeting it at L, then we can derive the following pair of parametric equations for the epicyclic curve:

$$x = r_1 cos\ \theta + r_2 cos\ k\theta$$

$$y = r_1 sin\ \theta + r_2 sin\ k\theta.$$

The angle $\psi = PON$ is the angle that gives the apparent position of the planet at P as seen from the center of the world O. Those positions of P for which $d\psi / dt = 0$ are called *stations*, from the Latin *stare, to stand still,* because the planet seems to stop in its orbit at those points. For those positions for which $d\psi/dt < 0$, the planet is said to be *in retrogression,* from the Latin *retrogradi, to go back,* since an observer at O will see it travelling backwards at such times. The angles Ψ and θ are related by the equation

$$\tan \Psi = y/x = [r_1 \sin \theta + r_2 \sin k\theta]/[r_1 \cos \theta + r_2 \cos k\theta]. \quad (*)$$

It is therefore possible to find those values of t, θ, or Ψ for which there are stations or retrogression by differentiating equation (*) with respect to t and solving $d\psi/dt = 0$ or $d\psi/dt < 0$.

As rich as the family of epicyclic curves is, Ptolemy found it necessary, in order to explain the appearances, to extend this class *a)* by introducing the device of *equants* and *b)* by removing the earth from the center of the world. The equant E_q is a point with coordinates $(-a,0)$, $r_1 > a > 0$, such that the angle $\omega = QE_qO$ rather than θ increases at a constant rate. The earth E_a is removed from the center of the world O to the point with coordinates $(a,0)$ collinear with E_q and O such that $E_q O = O E_a$, an equal distance that we have denoted by a. The planet P moves uniformly on the epicycle, whose center Q moves along the deferent so that ω increases uniformly. The regularity assumption now means that there are constants c_1 and c_2 such that $\omega = c_1 t$ and $\varphi = c_2 t$, so $\varphi = c_2 \omega / c_1$. If, finally, we put $c = c_2/c_1$, we have that the parametric equations for the path of P are

$$x = r_1 \cos \theta + r_2 \cos c\omega \text{ and } y = r_1 \sin \theta + r_2 \sin c\omega$$

where θ and ω are related by

$$\tan \omega = r_1 \sin \theta / [(r_1 \cos \theta) + a].$$

The stations and regressions are now determined according to the sign of $d\psi/dt$ where

$$\tan \psi = [r_1 \sin \theta + r_2 \sin c\omega]/[(r_1 \cos \theta + r_2 \cos c\omega) - a].$$

By adjusting r_1, r_2, and c, the epicyclic curves can be adjusted to take on all sorts of shapes, so that one of them could always be found to fit the trajectory of any body in the solar system. Only after the development of better instruments in the time of Kepler was it clear that the epicyclic curves did not give the exact paths of the planets, but the errors were nevertheless small.

epicycloid This word is composed of three Greek parts: the preposition ἐπί, which means *on*; the noun κύκλος, which means *circle*; and the suffix *-oid*, the metamorphosis of the noun εἶδος, *shape*. Like the related *hypocycloid*, it is due to Römer (1644–1710). A circle of radius b rolls without slipping on a circle of radius a. The trajectory of a fixed point P on the rolling circle is an *epicycloid*. In Euler's Latin, the curve is *epicyclois, epicycloidis*. The parametric equations of the epicycloid are

$$x = (a + b)cos\ \theta -\ b\ cos[(a + b)/b]\theta$$

$$y = (a + b)sin\ \theta -\ b\ sin[(a + b)/b]\theta.$$

If $a = nb$, the epicycloid has n cusps. If $n = 1$, Castillon (1704–1791) called the epicycloid the *cardioid*. If $n = 2$, it is the *nephroid*. The epicycloid of three cusps is called the *epicycloid of Cremona* after Luigi Cremona (1830–1903), who noticed it in his *Introduzione ad una Teoria Geometrica delle Curve Plane*, upon which his reputation rests.

epimorphism The Greek preposition ἐπί means *on*. The noun μορφισμός, *a shaping or forming*, is derived from the noun μορφή, *a form, shape*. An epimorphism is a homomorphism that is a surjection.

epitrochoid This word is composed of three Greek parts: the preposition ἐπί, which means *on*; the noun τρόχος, which means *a wheel*; and the suffix *-oid*, the transformation of the noun εἶδος, *shape*. If a circle of radius b rolls without slipping on a fixed circle of radius a, and if a fixed point P is at a distance c from the center of the rolling circle, then the locus of P is called an *epitrochoid*. If $c < b$, the curve is called a *curtate epitrochoid* (from the Latin *curtatus*, which means *shortened, reduced*). If $c > b$, the curve is called a *prolate epicycloid* (from the Latin *prolatus*, which means *brought forward, extended*). Both names are examples of a Latin adjective and a Greek-based noun happily married in a mathematical phrase. The parametric equations of the epicycloid are

$$x = (a + b)\cos \theta - c \cos[(a + b)/b]\theta$$

$$y = (a + b)\sin \theta - c \sin[(a + b)/b]\theta.$$

The prolate epicycloids have loops, which are finite in number if a and b are commensurable; if $a = nb$, there are n loops. If $n = 1$, the epitrochoid is the *limaçon* of Pascal. Lawrence says (p. 160) that the epitrochoids were first studied by Dürer in 1525.

epsilon *Epsilon*, the fifth letter of the Greek alphabet, is used in mathematics in the lower case (ε) to designate an arbitrarily small positive real number. The *epsilon-delta* definition of limits is due to Weierstraß.

equal The Latin adjective *aequus, -a, -um* means *plane, even, level*. From it was derived the verb *aequo, aequare, aequavi, aequatus* with the meaning *to make level, to make one thing equal to another*. The English adjective *equal* is derived from the Latin adjective *aequalis*, which was produced by adding the adjectival suffix *-alis* to the stem of the first principal part of *aequo*. The meaning of *aequalis* is *that which can be put on an equality with something else*. The diphthong *ae* was regularly shortened to *e* in the medieval period.

equant This adjective and noun, like *octant*, was formed on the analogy of *quadrant*. See the entry **quadrant**. The Latin adjective *aequus* means *level*, and the verb *aequo, aequare, aequavi, aequatus* means *to make level or equal*. Its present participle *aequans, aequantis*, whose stem is the English word, means *making level, equal, uniform, or regular*. The *punctum equans* or *regularizing point* in the Ptolemaic system is the point within the deferent circle such that the radius vector from this point to the center of the epicycle sweeps out equal angles in equal times. See the entry **epicycle**. When the equant point was adopted into the theory, the deferent circle was also referred to as the *circulus aequans* or *equant circle*; the equant point is not the center of the equant circle. The *equant point* and *equal circle* were needed in the Ptolemaic system in order to reconcile the actual celestial movements of the planets with the metaphysical assumption that all velocities in the heavens must be uniform.

equation The Latin verb *aequo, aequare, aequavi, aequatus* means *to make level or equal* and is derived from the adjective *aequus*, which means *level, equal*. The noun *aequatio, aequationis* meaning *a making equal* was formed from the fourth principal part, and our English word is the stem of this Latin noun. The work of the Venetian Count Riccati (1676–1754), who excelled in the science of acoustics, required him to consider the equation $y' + u(x)y^2 + v(x)y + w(x) = 0$, which is named *Riccati's equation* after him. The equations $y' + u(x)y = t(x)y^n$ are called *Bernoulli's equations* after Jakob Bernoulli, who first introduced them. His brother Johann showed that they may be solved by the substitution $z = y^{1-n}$, which reduces them to a first-order linear equation. The equation $x^2 y'' \, ax \, y' + b \, y = 0$ is called *Euler's equation* because it was solved by that mathematician during his work on the calculus of variations.

equator This Latin word means *that which makes equal*. The Latin adjective *aequus* means *level, equal*, and the verb *aequo, aequare, aequavi, aequatus* means *to make level or equal*. The celestial equator was so-called because when the sun is on it, day and night are of equal length; this great circle of the celestial sphere was the *equator diei et noctis*, the equalizer of day and night.

equi- This prefix gives the notion of equality to the adjective or noun that follows. It is the Latin *aequi-* from the adjective *aequus, -a, -um*, which means *equal*. *Aequilibritas* and *aequinoctium* are good Latin words. To write *equi-* for *aequi-* is wrong, though sanctioned by tradition, since *equi-* comes from *equus*, which means *horse*. The substitution of *e* for *ae* became common after the diphthong *æ* began to be pronounced as *ē*. The Greek equivalent of *equi-* is ἰσο-.

equiangular This adjective is composed of the Latin prefix *equi-* from *aequus, level,* and the adjective *angularis,* which means *pertaining to (-aris) an angle (angulus).* The *equiangular spiral* of Jakob Bernoulli is the polar curve with equation $r = ae^{b\theta}$. It is defined by the fact that the angle ψ between the radius vector and the tangent line is the same at all points. It is also called the *logarithmic spiral,* for obvious reasons. A sculptor was instructed to engrave it on the tombstone of Bernoulli in the cathedral of Basel, but he erred and made an Archimedean spiral instead. The inscription *Resurgo eadem mutata,* "I rise again, the same though changed," is a pun; the mathematical reference is to the fact that the pedal curve, the involute, and the evolute of the equiangular spiral are congruent equiangular spirals. Bernoulli referred to this spiral as *spira mirabilis,* the wonderful spiral. Torricelli proved that the length of that piece of the equiangular spiral with equation $r = ae^{b\theta}$ that begins at the point *(a,0)* and winds inward towards the pole is the same as the length of that piece of the tangent line to the spiral at *(a,0)* cut off between *(a,0)* and the line $\theta = 3\pi/2$. The equiangular spiral is an important pursuit curve that occurred in the work of a team of American mathematicians led by Leonard Gilman, who advised the U.S. government on strategies of pursuing U-boats during the Second World War. The dinner service of the Empress Elizabeth Petrovna consisted of porcelain plates decorated with a trellis of equiangular spirals in gold and magenta; reproductions were sold by the New York Metropolitan Museum of Art for $95 apiece.

equicontinuous This adjective is composed of the Latin prefix *equi-* from *aequus, level,* and the adjective *continuus,* which means *holding together.* Royden (p. 177) defines a family $\mathcal{F}$ of functions from a topological space X to a metric space $\{Y, \sigma\}$ to be *equicontinuous* at the point $x \in X$ if given $\varepsilon > 0$ there is an open set O containing x such that $\sigma[f(x),f(y)] < \varepsilon$ for all y in O and all $f \in \mathcal{F}$.

equidistant This adjective is composed of the Latin prefix *equi-* from *aequus, level,* and the adjective *distans, distantis,* which means *standing away from.* This word was used by the medieval Latin translators of

Euclid for the Greek παράλληλος, *parallel*. Euclid himself actually rejected the temptation to define *parallel* using the notion of equidistance; rather, he based his theory on the notion of non-secancy.

equidistribution See the entries **equi-** and **distribution**. Let 1_A be the characteristic function of the set A, and let $[x]$ be the integer part of x. Put $\langle x \rangle = x - [x]$. The *equidistribution theorem of Hermann Weyl* (1916) says that if x is an irrational number and A is a measurable subset of $[0,1]$, then $\{1_A(x) + 1_A(2x) + 1_A(3x) + \cdots + 1_A(nx)\}/n$ tends to the measure of A as n tends to infinity. V. I. Arnold communicated to A. Avez a problem of Gelfand that is a corollary to Weyl's theorem. Consider the sequence of first digits of the numbers m^n, where m is a positive integer that is not an integral power of 10, and $n = 1, 2, 3, \ldots$. For $k = 1, 2, 3, 4, 5, 6, 7, 8, 9$, let $f_k(n)$ be the number of k's among the initial n elements of the sequence of first digits. Gelfand asked for the limiting ratio of k's, that is, he asked if $F_k = \lim f_k(n)/n$ exists as n goes to infinity, and if so, what it is. The limiting ratio of k's exists and is indeed independent of m; in fact $F_k = log_{10}\{(k + 1)/k\}$. Thus we have, to five decimal places:

$$F_1 = .30103$$
$$F_2 = .17609$$
$$F_3 = .12494$$
$$F_4 = .09691$$
$$F_5 = .07918$$
$$F_6 = .06695$$
$$F_7 = .05799$$
$$F_8 = .05115$$
$$F_9 = .04576$$

Why is this sequence monotonically decreasing? For a proof, see "On the Origin and History of Ergodic Theory" by A. Lo Bello, *Bollettino di Storia delle Scienze Matematiche*, vol. I, no. 1, 1983, pages 37–75.

equilateral This adjective means *pertaining to a side*, and is composed of the Latin prefix *equi-* from *aequus*, *level*, and the adjective *lateralis*, which is formed by adding the adjectival suffix *-alis* to the stem of the noun *latus, lateris, side*.

equilibrium This noun is composed of the Latin prefix *equi-* from *aequus, level*, and the noun *libra*, which means *a scale*. The noun *aequilibrium* is medieval Latin.

equipollent This adjective is composed of the Latin prefix *equi-* from *aequus, level*, and the present participle *pollens, pollentis*, of the verb *polleo, pollere*, which means *to be strong, mighty, powerful*. The participle *aequipollens, aequipollentis* therefore means *being equally powerful*.

equipotent This adjective is composed of the Latin prefix *equi-* from *aequus, level*, and the present participle *potens, potentis*, of the verb *possum, posse, potui*, which means *to be able, mighty, powerful*. The participle *aequipotens, aequipotentis* therefore means *to be equally capable or powerful*.

equipotential This word is formed by adding the Latin adjectival suffix *-alis* to the stem of the noun *aequipotentia*, which means *equal power*. It is therefore an adjective meaning *having equal power*.

equiprobability An *equiprobability space* is a probability space whose sample space is a finite set $\{x_1,...,x_n\}$ and whose probability measure assigns the same value $1/n$ to each outcome. It is the simplest situation to arise, and the examination of it is ascribed to Laplace (1749–1827). In fact, the assignment of equal probabilities is called after him the *Laplace definition of probability*.

equivalence This is the medieval Latin noun *aequivalentia*, formed from the verb *aequivaleo*. See the following entry.

equivalent This adjective is derived from the present participle *aequivalens, aequivalentis* of the verb *aequivaleo, aequivalere*, which means *to be equally strong*. The verb is composed of the Latin prefix *equi-* from *aequus, level*, and the present participle *valens, valentis*, of the verb *valeo, valere, valui, valitus*, which means *to be strong*.

ergodic This word was coined by Boltzmann from the Greek ἔργον, *work*, and ὁδός, *path*. He used it for his hypothesis that each surface of constant energy consists of a single trajectory. It is first to be found on page 201 of his paper "Über die mechanischen Analogien des zweiten Hauptsatzes der Thermodynamik," *Journal für die reine und angewandte Mathematik*, 100 (1887), pages 201–212.

error This is the Latin word for *mistake*. It is derived from the verb *erro, errare, erravi, erratus*, which means *to wander*.

escribed This is a low word, the product of ignorance. The Latin verb *exscribo, exscribere, exscripsi, exscriptus* means *to write out*; it is the compound of the preposition *ex, out*, and the verb *scribo, to write*. The force of the prefix *ex-* is to emphasize the thoroughness of the writing, not its location. *To write outside* would have been *extra scribere*. There is no Latin verb *escribo*. An *escribed circle* of a triangle is a circle that is tangent to one side of a triangle and to the extensions of the other two sides; there are three such circles of every triangle. The use of the English verb *escribe* in this mathematical sense (rather than just *to write out or copy out*) is of the nineteenth century.

essential The noun *essentia*, which means *being*, is derived from the infinitive *esse* of the Latin verb *sum, esse, fui, futurus*, which means *to be*. By the addition of the adjectival suffix *-alis* there was produced the adjective *essentialis*, whence the English adjective is derived by removal of the nominative case ending *-is*. The Latin noun was invented to translate the Greek noun οὐσία, *being*, when there was a need for a Latin edition of the Platonic dialogues.

estimate This word is derived from the fourth principal part of the verb *aestimo, aestimare, aestimavi, aestimatus*. See the following entry.

estimator The Latin verb *aestimo, aestimare, aestimavi, aestimatus* means *to evaluate, consider*. From its fourth principal part, upon the addition of the suffix *-or* to the stem, proceeds the noun of agent *aestimator* with the meaning *he who evaluates*.

ethnomathematics This new word is correctly formed from the Greek ἔθνος, *nation, people,* and τὰ μαθηματικά, *mathematics. Ethnomathematics* is the name of a new subject. Just as the plural τὰ ἔθνη of τὸ ἔθνος means *the nations* in the exclusive sense of *the gentiles,* that is, everyone except the chosen people, so ethnomathematics means the mathematics of all primitive or ancient peoples except those who actually developed the subject.

eu- The prefix *eu-* is the Greek εὖ-, which means *well.* The opposite suffix is *dys-,* corresponding to our *un-* and *mis-,* from the Greek δυς-, an inseparable prefix.

Euclid The name Euclid is the Greek Εὐκλείδης and means *beautifully named, well-called.* Everything known about Euclid of Alexandria is contained in a passage of the commentary of Proclus on the first book of the *Elements of Geometry:*

> Not long after these men came Euclid, who brought together the *Elements,* systematizing many of the theorems of Eudoxus, perfecting many of those of Theaetetus, and putting in irrefutable demonstrable form propositions that had been rather loosely established by his predecessors. He lived in the time of Ptolemy the First, for Archimedes, who lived after the time of the first Ptolemy, mentions Euclid. It is also reported that Ptolemy once asked Euclid if there was not a shorter road to geometry than through the *Elements,* and Euclid replied that there was no royal road to geometry. He was therefore later than Plato's group but earlier than Eratosthenes and Archimedes, for these two men were contemporaries, as Eratosthenes somewhere says. Euclid belonged to the persuasion of Plato and was at home in this philosophy, and this is why he thought the goal of the *Elements* as a whole to be the construction of the so-called Platonic figures. (Proclus, *A Commentary on the First Book of Euclid's Elements, translated, with introduction and notes,* by Glenn R. Morrow, Princeton University Press, 1970, pp. 56–57)

All else about the *curriculum vitae* of Euclid (except, of course, for his works) is speculation and romance. As for the episode of the "royal road," if it is not true, it is certainly, as the Italians say, a good story.

Since Ptolemy I Soter reigned in Egypt from 323 to 285 B.C., and Proclus lived from A.D. 410 to 485, there is a gulf of three-quarters of a millennium between the subject of the paragraph just quoted and its author. For Proclus the reader may consult with profit the book *What is Ancient Philosophy?* by Pierre Hadot, published in English translation by Harvard University Press in 2002.

In an age before Xerox machines, those who transcribed manuscripts made the usual copying mistakes, and changed the text where they thought it recommendable to do so. So, for example, Proposition 12 of Book III, that if two circles are tangent externally, their centers and the point of tangency are collinear, is an addition to Euclid of a theorem of Heron. Among those who modified the *Elements* of Euclid was Theon the mathematician, the father of Hypatia. He flourished, or rather taught mathematics, at Alexandria in the second half of the fourth century A.D. The edition of the *Elements* that he produced was so wonderful that it displaced all others. For 1,300 years, the only known manuscripts of the Greek *Elements* were those that boasted to be from the edition, or from the lectures, of Theon, and others that were clearly dependent upon them. The human race progressed to the point where men demanded to have the text of the *Elements* exactly as it had left Euclid's hands, or as close to that state as possible. Fortunately for them, Theon, in his commentary on Ptolemy's *Almagest*, had congratulated himself for an improvement that he had made to Proposition 33 of Book VI:

> In equal circles, angles have the same ratio as the circumferences on which they stand, whether they stand at the centers or at the circumferences.

Theon was proud of having added at the end of this statement "and further the sectors, as constructed from the centers," and of having extended the proof accordingly. The problem, therefore, of finding a text of the *Elements* unmolested by Theon was reduced to that of finding a manuscript without the aforementioned addition in the sixth book. Such a manuscript was found in Paris in 1808 by F. Peyrard (1760–1822), who was examining manuscripts stolen by Napoleon from the Vatican and shipped to France. This manuscript, the most authoritative of all manuscripts of the *Elements*, was returned to the Holy See by Louis XVIII in 1814; it is now called *P*, or more officially *Vat. Gr. 190*. It is of the ninth century, parchment, now in two volumes quarto (390 x 235 mm). The critical edition of the Greek text of the *Elements*, based on *P*, was published by J. L. Heiberg (1854–1928) between 1883 and 1888. An English translation of Heiberg's text, with introduction and commentary, was the masterpiece of Thomas L. Heath (1861–1940) and appeared in 1908. Heath subsequently published an edition of the Greek text of Book I with notes for the use of students of mathematics who could read Greek, a type of scholar that does not exist anymore.

In modern times, Hilbert corrected the deficiencies in the logical structure of the *Elements* and thereby brought the Euclidean geometry up to the modern standards of rigor. In 2000 Hartshorne published *Geometry: Euclid and Beyond* at Springer-Verlag, in which he shows with all details how the propositions of Euclid are deduced from the axiom system of Hilbert.

Most of this entry was taken, with minor changes, from my volume *The Commentary of al-Nayrizi on Book I of Euclid's Elements of Geometry, With an Introduction on the Transmission of Euclid's Elements in the Middle Ages*, Brill Academic Publishers, Inc., Boston and Leiden, 2003, pages 1–3.

Euclidean This is the nominal adjective meaning *pertaining to Euclid*. The word was used by Barrow in the seventeenth century. To write *euclidean* with a small *e* is not a practice of the best authors and is therefore not correct.

eureka This is the exclamation of Archimedes, εὑρηκα—*I have found it!*—which he uttered (so says Plutarch) upon discovering the principle of hydrodynamics.

evaluate The Latin verb *evalesco, evalescere, evalui* means *to grow strong*. It is derived from the preposition *e* and the verb *evaleo, evalere, evalui*, which means *to grow strong*, which is itself derived from the verb *valeo, valere, valui*, which means *to be strong*. At some point a first-declension transitive verb *evaluo, evaluare, evaluavi, evaluatus* came into usage with the meaning *to find the value of.* The verb *evaluate* and the noun *evaluation* are both derived from the fourth principal part of this later verb; the latter came over from the French language, as if there were a Latin noun *evaluatio, evaluationis.*

[teaching] evaluation The custom of administering teaching evaluations at American colleges is universal. If Socrates or Jesus Christ appeared on a campus, they would be subjected to such a process. It is not necessary to make a case *for* teaching evaluations by students because everyone is doing it. The philosophical case *against* teaching evaluations by students was made by Deming in *The American Statistician*, February 1972, page 47:

Dear Sir,

Memorandum on Teaching

There is much discussion today about student participation in affairs of the university, even in respect to evaluation of teachers and content of courses. Here are some of my thoughts on the matter.

It seems to me that the prime requirement for a teacher is to possess some knowledge to teach. He who does no research possesses not [*sic*] knowledge and has nothing to teach. Of course some people that do good research are also good teachers. This is a fine combination, and one to be thankful for, but not to expect.

No luster of personality can atone for teaching error instead of truth. One of the finest teachers that I ever knew could hold 300 students spellbound, teaching what is wrong. The two poorest teachers that I ever had (though a third one ran neck and neck) were Professor Ernest Brown in mathematics at Yale and Sir Ronald Fisher at University College in London. Sir Ernest will be known for centuries for his work in lunar theory, and Sir Ronald for revolutionizing man's methods of inference. People came from all over the world to listen to their impossible teaching and to learn from them, and learn they did. I would not trade my good luck to have had these men as teachers for hundreds of lectures by lesser men but "good teachers."

It is too late when the student finds out that the foundation that he built in college is shaky. He may fill in gaps by self-study, but the place to lay the sure foundation is in school. The best insurance that a student can take out is to make sure that his teachers do research.

The student is at a disadvantage when asked to evaluate a teacher. On what basis? Content of course? Lustre of personality? Knowledge of the subject? The teacher's interest in making sure that he is communicating to the students whatever it is that he is trying to say? A student can possibly judge the teacher's knowledge of the subject, but he can hardly be a judge of the content of the course. Not even the teacher has dependable knowledge about what ought to be taught. Learning today is preparation for 5, 10, 20 years in the future. A student naturally likes what he calls a good teacher, for whatever reasons. What use, then, could be made of students' evaluation of a teacher?

The problem of identifying a good teacher is not one in consumer research, though every statistician knows well the importance of consumer research. A university should be now, as in days gone by, a place where one may listen and learn from great men.

The only suitable judges of a teacher's knowledge are his peers. The only objective criterion of knowledge is research worthy of publication. Publication should of course be measured on some scale of contribution to knowledge, not by number of papers.

Suggestions from students concerning the content of a course or the competence of a teacher are accordingly, in my judgment, a reckless idea. I would counsel myself and my colleagues to take no notice of evaluation by students. For my own part, I could not teach under a system of evaluation by students.

> W. Edwards Deming
> Professor of Statistics
> Graduate School of Business Administration
> New York University

Since mathematicians ought not to participate in nonsense, it is necessary to inquire into the scientific content of any method of teaching evaluations, to determine if that method contradicts knowledge.

On August 10, 1998, I sent a description of the "Teaching Evaluation Instrument" (not composed by me) used at a typical college to several dozen American statisticians:

For each instructor in each course, the average score on a question is calculated, and this score is compared with all the other average scores on the same question of all other instructors in all other courses in the College as follows, by what the authors call *The Box and Whiskers Method.* One first finds the twenty-fifth percentile, the median, and the seventy-fifth percentile of all the mean scores of all the instructors in the College on the given question. One then draws a box above the number line from the twenty-fifth percentile to the seventy-fifth percentile, and marks the College-wide median in that box with a vertical line, *viz.* |. One then draws whiskers on both sides of the box, each whisker of length equal to the length of the box. The faculty member who wants to interpret his score plots it with an *x* on the diagram. If the score falls within the box, it is average. If it falls on a whisker, he is supposed to interpret his score as above or below average depending on whether the *x* falls on the right or the left whisker. If the score falls to the right of the right whisker, his score is way above average; if it falls to the left of the left whisker, his score is way below average.

It will be immediately evident to those competent to have an opinion about descriptive statistics that the authors of the instrument have tinkered with the true "box plot" method and used it for a purpose for which it was not intended.

The following is an excerpt from a personal letter to me from John E. Freund in reply to my message about the "Teaching Evaluation Instrument" described above:

August 14, 1998

Dear Professor Lo Bello:

First let me explain a few things about box-and-whiskers plots. They were introduced by John Tukey as part of his contribution to what is nowadays called EDA; namely, Exploratory Data Analysis. The name reflects his somewhat whimsical approach to things and most of us refer to them nowadays simply as box plots. This is the term I use in my elementary books.

...The whisker on the left extends from the smallest value in a set of data to Q_1, the first quartile, which, of course, is the same as the 25th percentile. The whisker on the right extends from Q_3, the third quartile, (same as 75th percentile) to the largest value.

Box plots were introduced mainly to describe the "shape" of relatively small sets of data. When you have lots of data, you can group them and draw a histogram, which displays the general shape (symmetry or skewness) of a set of data. This is not the case when you have relatively small sets of data, consisting, say, of 10 or 20 measurements.

With a box plot you can judge the shape of a small set of data by observing whether the median is in the middle of the box (indicating symmetry) or whether it is closer to Q_1 or Q_3 (indicating skewness). The whiskers also serve for this purpose. A long whisker on the right is an indication of positive skewness and a long whisker on the left is a sign of negative skewness. In my opinion, this is about the only reason for using box plots— judging the overall shape of relatively small sets of data. Unless this is of some special purpose in studying your teacher evaluations, I can't see any reason for using box plots in your problem. You might say that box plots are somewhat of a fad— showing that you are really on the ball (?)—but that would hardly justify their use in connection with your situation.

So, I wouldn't say that box plots are mathematical nonsense. Their use is appropriate in some special situations, but I can't see much use for box plots in comparing your ratings of teachers.

Letting students evaluate teachers by means of questionnaires is a popular pastime, of which I have never been very fond. Personally, I <u>know</u> whether or not I am a good teacher, regardless of what some students may have to say on the subject. Such evaluations are terribly subjective and must be interpreted as such. Do you seriously think that any student to whom you gave an F will rate you an excellent teacher?

As I pointed out earlier, box plots serve only one purpose—to judge the symmetry or the lack of it for relatively small sets of data. If a set of data is more or less symmetrical, using standard units (or units of standard deviation, as you call it) makes much more sense when you want to compare individual figures with sets of data, but mainly if your whole distribution is fairly symmetrical. Otherwise you may have to resort to some sort of witchcraft. Please note that I pointed out earlier that the "whiskers" are generally not of equal length. It is the possible difference in their length that helps in judging the symmetry or the lack of it in a set of data. So, you might draw a box plot to judge whether a set of data is symmetrical, and if it seems to be symmetrical, throw out the box plot and use standard units. If you are dealing with a fairly substantial set of data, I would still prefer simply looking at a histogram...

Best regards,

John E. Freund

event The Latin verb *evenio, evenire, eveni, eventus* means *to come (venio) out of (e-)*. From its fourth principal part came the noun *eventus, eventūs* with the meaning *consequence*. *Event* is the name for an element of the sigma-field of those subsets of the sample space to which one assigns probabilities.

evolute The Latin verb *evolvo, evolvere, evolvi, evolutus* means *to roll out*. The *evolute* of a curve C is the locus of the center of curvature P to C at Q as Q varies over C. The definition is due to Huygens (1673), although Apollonius discusses evolutes in Book V of his *Conics*. If C has equation $y = f(x)$ and therefore parametric equations $x = t$ and $y = f(t)$, then the parametric equations of the evolute are

$$x = t - \{f'(t)[1 + (f'(t))^2]/f''(t)\}$$

$$y = f(t) + \{[1 + (f'(t))^2]/f''(t)\}.$$

The *Oxford English Dictionary* traces the first English appearance of the word *evolute* to the period 1730–1736.

exact The Latin verb *exigo, exigere, exegi, exactus* means *to drive (ago) out (ex-)*, *to do completely, demand, complete, finish*. The fourth principal part *exactus, -a, -um* came to mean *accurate, precise*. A differential equation $M(x,y) + N(x,y)\,y' = 0$ is *exact* if $N_x(x,y) = M_y(x,y)$. It is so-called because it was proven by Clairault (1713–1765) that this is precisely the condition for there to exist a function $z = f(x,y)$ such that $\nabla f(x,y) = (M(x,y), N(x,y))$; that is, M and N are exactly the partial derivatives of some function $z = f(x,y)$. The solutions are $f(x,y) = c$, that is, the level curves of the function $z = f(x,y)$.

example The Latin verb *eximo, eximere, exemi, exemptus* means *to take out, to take away*, compounded of *emo, emere, emi, emptus, to take*, and *ex, out of*. The related noun *exemplum* means *something chosen from a number of things, a sample*.

excenter This is a low word. It means the *center of an excircle of a triangle*. An intelligent observer who did not know this would guess that it means something taken out of the center. See the entry **escribed**.

excess The Latin verb *excedo, excedere, excedi, excessus* means *to go (cedo) out (ex), to pass beyond*. Its first principal part is the origin of the English word *excede*, and its fourth principal part has produced our word *excess*. The Romans also had a fourth-declension noun *excessus, excessūs* meaning *departure*.

excircle This is a very low word. One would imagine it to mean *something taken from a circle*, as *ex libris* means *something taken from the books*, but one would be wrong. See the entry **escribed**.

exclusionary The graduate alumni of the Yale University Department of Mathematics received in March 2004 a letter from Richard Beals addressed to "Fellow Alumnus/Alumna." The first paragraph was:

> Looking at last year's letter, I note that I owe an apology for the exclusionary salutation "Dear Fellow Alumnus." I wasn't thinking.

A *mea culpa* of this sort from one of the highest authorities of mathematics is an instructive episode. The usual device to which people resort in order to avoid the accusation of exclusion is the slash, pompously called the *virgule*. Its use is catastrophic for prose style. Paul Halmos addressed the issue of inclusive language in the preliminary matter to his "automathography" *I Want to Be a Mathematician* (MAA, 1985), where he inserted the following clarification in the preliminary matter:

My expository style relies heavily on the exemplary singular, and the construction "everybody...his" therefore comes up frequently. This "his" is generic, not gendered. "His or her" becomes clumsy with repetition and suggests that "his" alone elsewhere is masculine, which it isn't. "Her" alone draws attention to itself and distracts from the topic at hand. "Their" solves the problem neatly but substitutes another. "Ter" is bolder than I am ready for. "One's" defeats the purpose of the construction, which is meant to be vivid and particular. "Its" is too harsh a joke. Rather than play hob with the language, we feminists might adopt the position of pitying men for being forced to share their pronouns around.

The best advice in this regard is given by Alford (§381):

Avoid all oddity of expression. No one ever was a gainer by singularity in words, or in pronunciation. The truly wise man will so speak, that no one may observe how he speaks. A man may show great knowledge of chemistry by carrying about bladders of strange gases to breathe; but he will enjoy better health, and find more time for business, who lives on the common air. When I hear a person use a queer expression, or pronounce a name in reading differently from his neighbours, the habit always goes down, in my estimate of him, with a *minus sign* before it; it stands on the side of deficit, not of credit.

exhaustion The Latin verb *exhaurio, exhaurire, exhausi, exhaustus* means *to draw (haurio) out (ex), to drain out, to empty*. From the fourth principal part is derived the noun *exhaustio, exhaustionis* with the meaning, *a draining out*. The stem of this noun is the origin of the English word.

expand The Latin verb *expando, expandere, expandi, expansus* means *to stretch (pando) out (ex-), extend*.

expansion The late Latin noun *expansio, expansionis, a stretching out,* is derived from the fourth principal part of the verb *expando*. (See the preceding entry.)

expectation The Latin verb *exspecto, exspectare, exspectavi, exspectatus* means *to look for, to await.* From the fourth principal part of this verb there proceeded the noun *exspectatio, exspectationis* with the meaning *a waiting for, a looking for.*

expected value The English verb *expect* is the Latin verb *exspecto* with the *s* removed because of careless pronunciation. The expression *expected value* is a synonym for *mean* or *average value* of a random variable.

experiential From the Latin adjective *peritus, -a, -um,* which means *learned,* is derived the verb *experior, experiri, expertus sum* with the meaning *to put to the test.* From this verb proceeds the noun *experientia,* which means *a testing,* and from the noun comes the English adjective *experiential* meaning *pertaining to experience.* This adjective has now become cant.

explicit The Latin verb *plico, plicare, plicavi, plicatus* means *to fold.* The addition of the prefix *ex-* produces the verb *explico, explicare, explicavi, explicatus* meaning *to unfold, to disintangle, to unroll.* The final *i* is due to the fact that there were alternate third and fourth principal parts *explicui, explicitus,* the latter of which had the meaning *straightforward, easy.*

exponent The Latin verb *expono, exponere, exposui, expositus* means *to put (pono) out (ex-), to cast out. Exponent* is derived from the stem of its present participle *exponens, exponentis.*

exponential This is the stem of a Latin adjective *exponentialis* invented in the modern period by mathematicians writing in Latin. It was formed by adding the adjectival suffix *-ialis* to the stem of the present participle *exponens, exponentis* of the verb *expono.* See the preceding entry. The exponential distribution of probability explains the waiting time until the occurrence of a rare event.

extend The Latin verb *extendo, extendere, extendi, extensus* means *to stretch (tendo) out (ex-), expand.*

extension The Latin noun *extensio, extensionis* is derived from the fourth principal part of the verb *extendo.* See the preceding entry.

exterior This is the comparative degree of the Latin adjective *exter* (also *exterus*), which means *outward, foreign, strange.*

extouch See the article **excircle**. Each excircle of a triangle touches it at one and only one point. The *extouch triangle* is the triangle formed by connecting the three points where the triangle's three excircles touch it. *Ex* is a Latin preposition meaning *out of,* and *touch* is the stem of the French *toucher,* a word not derived from Latin. This ugly word was coined by someone with no taste at all. Furthermore, if *extouch* were to have any meaning at all, it would have to be *to touch out of,* not *to touch on the outside.*

extract The Latin verb *extraho, extrahere, extraxi, extractus* means *to drag (traho) out (ex-).*

extrapolate The Latin adverb *extra* means *outside*; the phrase *extra omnes* means *Everybody out!* The verb *polio, polire, polivi, politus* means *to polish, file, make smooth.* From these two words were formed the adjective *interpolis* with the meaning *furbished up, vamped up* and the verb *interpolo, interpolare, interpolavi, interpolatus* with the meaning *to furbish, vamp up,* and thereby *to falsify.* From the last principal part of the verb came the English word *interpolate.* On the analogy of this verb there was created in the nineteenth century the verb *extrapolate,* which did not exist previously, as if from a Latin verb *extrapolo, extrapolare, extrapolavi, extrapolatus,* which also did not exist. If a least-squares line is constructed on the basis of data points $(x_1,y_1),\ldots,(x_n,y_n)$, *to extrapolate* is to evaluate that line at a point outside of the smallest interval containing the x_i's.

extremal This word is formed by superimposing the adjectival suffix *-alis* on the stem of *extremus*, which is already an adjective; such a process may be expected to produce a low word. It means *having something to do with extremes* as in *extremal solutions*, solutions that are extreme in some sense.

extreme This is derived from the Latin adjective *extremus*, which is the superlative degree of *exter* (also *exterus*), which means *outward, foreign, strange*.

F

F The F-distribution of mathematical statistics is due to Snedecor (1881–1974) and named by him after Ronald Fisher (1890–1962).

face This is derived from the Latin noun *facies*, which means *face*.

facilitator This is the modern name for someone who presides over or organizes a session at a meeting; the intent is that the title of such a fellow should be as humble as possible. The Latin adjective *facilis* means *easy to do* and is derived from the verb *facio, facere, feci, factus, to do*. The overused modern word *facilitator* appears to be a noun of agent formed from the fourth principal part of a frequentative verb *facilito, facilitare, facilitavi, facilitatus* with the meaning *to keep on making easy*, but there is no such verb. The word is pure cant.

factor This is a Latin noun meaning *he who does, maker, creator*, from the verb *facio, facere, feci, factus, to do*. The factors of an integer are viewed as making the integer. The use of the word *factor* in the mathematical sense is traced by the *Oxford English Dictionary* back to 1673, where the definition appears in Kersey's *Algebra*. In English this word may be either a noun or a verb; it was the former first. It then became a verb as well at the hands of those with no literary taste. If the equation $M(x,y) + N(x,y)y' = 0$ is not exact, it can be made exact by multiplying through by a suitable function $u(x,y)$. Such a function u is called an *integrating factor*. If the quotient $(M_y - N_x)/N$ contains no y, then there is an integrating factor

$$u = exp \int (M_y - N_x)/N dx.$$

Similarly, if there is an integrating factor depending solely on x, it is given by

$$u = exp \int (N_x - M_y)/M dy.$$

factorable This is a word derived from the modern Latin *factorabilis*, *capable of being factored*, as if there were a Latin verb *factoro, factorare* meaning *to factor*. The suffix *-abilis* is the origin of our word *able* and is derived from the adjective *habilis*, *capable*. It is added to the verb for the activity of which one is supposed to be capable.

factorial This word was created by adding the Latin suffix *-ialis* to the stem *factor* in order to make an adjective out of the noun. It was formed incorrectly since there is no reason for the connecting vowel *i*; the word should have been *factoralis*. *Factorial* with the *i* should really mean *having to do with a factory*. The common notation *n!* is read "*n factorial*" and is defined to be $1 \cdot 2 \cdot 3 \cdot 4 \cdots n$ for every positive integer *n*. In order to sustain a pattern evident in various formulas, $0!$ is defined to be 1. Schwartzman writes *sub voce*, "The exclamation mark was first used to represent factorials in 1808 by Christian Kramp (1760–1826)." The history of the various notations used to represent $1 \cdot 2 \cdot 3 \cdot 4 \cdots n$ is told in two fascinating paragraphs of Cajori's *A History of Mathematical Notations*, §§448–449. The Germans read *n!* as *n Facultät*. The exclamation mark eventually replaced the notation ∟ that Isaac Todhunter had made popular in Britain and America; the *n* was written inside the ∟. On this change Augustus De Morgan wrote in 1842:

> Among the worst of barbarisms is that of introducing symbols which are quite new in mathematical, but perfectly understood in

common, language. Writers have borrowed from the Germans the abbreviation *n!* to signify *1·2·3·4···(n − 1)n*, which gives their pages the appearance of expressing surprise and admiration that 2, 3, 4, etc., should be found in mathematical results.… (Quoted by Cajori, Florian, in *A History of Mathematical Notations*, Dover Publications, Inc., two volumes bound as one, New York, 1993, §713)

factorization The English verb *to factorize*, is obsolete; nevertheless, the noun *factorization* formed from it has survived. See the preceding entry. The Greek ending -ιζειν has been added to a Latin noun to produce a make-believe verb *factorizo*; from what would have been the fourth principal part *factorizatus* of this creation was produced the noun *factorization* on the usual model.

fallacy *Fallax, fallacis* means *false* in Latin; the noun *fallacia* means *deceit, trick, fraud*. They are both derived from the verb *fallo, to deceive*.

false This adjective is derived from the fourth principal part of the verb *fallo, fallere, fefelli, falsus, to deceive*.

family This noun is taken from the Latin *familia, family*.

figurate number *Figura* is the Latin word for *form, shape, figure, size*. From this noun came the verb *figuro, figurare, figuravi, figuratus, to form, mould, or shape*. *Figurate* is from the fourth principal part of this verb. Triangular (1, 3, 6, 10,…), square (1, 4, 9, 16,…), and cubic (1, 8, 27, 64,…) numbers are examples of such figurate numbers.

figure See the previous entry. The corresponding Arabic word شاكل was used by metonomy for *theorem*, and this carried over into Latin, where the early translators used *figura* and spoke, for example, of "the first figure of the first book."

filter The medieval Latin word for the material *felt* was *filtrum*. If $\mathcal{B}$ is a non-empty collection of subsets of a set X, then $\mathcal{B}$ is a *filter* for X if 1) $A, B \in \mathcal{B} \Rightarrow A \cap B \in \mathcal{B}$ and 2) $A \in \mathcal{B}$ and $B \subseteq X \Rightarrow A \cup B \in \mathcal{B}$.

finite The Latin verb *finio, finire, finivi, finitus* means *to set boundaries, to enclose within limits*. It is a denominative verb from the noun *finis*, which means *boundary*. The English adjective was created from the fourth principal part.

fixed The Latin verb *figo, figere, fixi, fixus* means *to fix, fasten*. From its fourth principal part came the English verb with the same meaning *to fix*.

fluxion The Latin word *fluo, fluere, fluxi, fluxus* means *to flow*. The associated noun *fluxus, fluxūs* means *a flowing*, and the past participle *fluxus, -a, -um* means *leaky*. The word *fluxio, fluxionis* did not exist (except as an error for *fluctio*) until it was created by Newton. The Romans, however, had nouns *fluctio, fluctionis* and *fluctus, fluctūs*, both meaning *a flowing*.

focal From the Latin noun *focus, a fireplace*, there was formed the adjective *focalis* meaning *pertaining to the fireplace*, whence the English adjective was derived by dropping the nominative ending *-is*.

focus See the preceding entry. The plural must be the Latin *foci*. To write *focuses* is not good style since it is done only by inadequately educated authors.

folium This is the Latin word for *leaf*. The Latin plural *folia* must be used; the absurd *foliums* is not an option.

folium Cartesianum The name means the *leaf of Descartes*. This plane curve with loop and asymptote was invented by Descartes; its parametric equations are

$$x = 3at/(1 + t^3), \qquad y = 3at^2/(1 + t^3);$$

elimination of the parameter gives the equation $x^3 + y^3 = 3axy$. The polar equation is

$$r = (3a \sin \theta \, \cos \theta)/(\sin^3 \theta + \cos^3 \theta).$$

It is remarkable in that the area enclosed by the loop is equal to the area intercepted between the folium and its asymptote, $3a^2/2$. (The trisectrix of Maclaurin also has this property.) The equation of the asymptote is $x + y + a = 0$.

form This is from the Latin noun *forma*, which means *figure*, *shape* and eventually *a nice figure, a nice shape, beauty*. The Roman translators of Plato used it to translate the Greek ἰδέα; for this reason the Platonic *ideas* are usually called in English the Platonic *forms*.

formula In Latin, this is the diminutive of *forma* and means *a little figure* or *shape*, and then *physical beauty*. Euler's formula is the identity $e^{i\theta} = \cos \theta + i \sin \theta$, from which we get the identities $e^{i\pi} = -1$ and $e^{-\pi/2} = i$.

fractal The Latin verb *frango, frangere, fregi, fractus* means *to break*. The addition of the suffix *-al* to the stem of the fourth principal part produced the English word *fractal* as if copied from a non-existent Latin *fractalis*.

fraction The Latin noun *fractio, fractionis, breaking*, was derived from the fourth principal part of the verb *frango* by adding the nominal ending *-io*.

frequency The Latin adjective *frequens, frequentis* means *crowded*. The noun *frequentia, a big assembly, an increase in population, a large number*, was formed by adding the nominal ending *-ia* to the adjective's stem. It became English by dropping the nominative ending *-a* and changing the final *i* to the fancier *y*.

friction The Latin verb *frico, fricare, fricui, frictus* means *to rub*. From the stem of the fourth principal part there was created the noun *frictio, frictionis*, from whose stem there was created the English noun *friction*.

frustum This is the Latin word for *a piece, a bit, a morsel*. It is related to the deponent verb *furor, frui, fructus sum, to enjoy*. For the plural, either *frusta* (the Latin plural) or *frustums* is permissible. The frustum of a pyramid is the prismatoid that remains when a smaller pyramid is lopped off the top of a larger one by slicing parallel to the base. If the frustum has altitude h and square bases of side b on top and of side B on the bottom, then its volume is $h(b^2 + bB + B^2)/3$, a formula that was known to the author of the Moscow Papyrus (problem 14). Furthermore, if in this case we put $r = b/B$, then the centroid of the frustum is at a height $h(1 + 2r + 3r^2)/4(1 + r + r^2)$ above the base. A pyramid formed of successively smaller frusta placed one on top of another is a *step pyramid*; the most famous of these is the step pyramid of Pharaoh Zoser at Saqqara, which consists of six frusta and is 201 feet tall. One can use the data given in *Baedeker's Lower Egypt* (London, 1885) to find that its centroid is 79 feet up from the ground.

function The deponent verb *fungor, fungi, functus sum* means *to occupy oneself with anything, to perform, to discharge [an office]*. From its fourth principal part there was formed the noun *functio, functionis* meaning *that which is performed or discharged*, by the addition of the nominal ending *-io*. The English noun comes from the stem of the Latin parent.

functional This was originally an adjective formed by adding the suffix *-alis* to the stem of the noun *functio, functionis, a performance, an executing*, thereby producing the adjective *functionalis* meaning *pertaining to a function*. When it came into English, the nominative singular case ending *-is* was dropped.

fundamental The Latin word *fundus, -ī* means *the ground*. From it was formed the verb *fundo, fundare, fundavi, fundatus* with the meaning *to found*. From this verb, by the addition of the suffix *-mentum*, is formed the result of the founding, the *fundamentum*, the *foundation*. The adjective *fundamentalis, pertaining to the foundation*, is formed by adding the suffix *-alis* to the stem of the noun. The nominative ending *-is* was then dropped, as usual, when the word came into English.

G

γ *Gamma* is the third letter of the Greek alphabet. The area of the plane region H_n bounded by the *x-axis*, the hyperbola with equation $y = 1/x$, and the vertical lines $x = n$ and $x = n + 1$ is $ln[(n + 1)/n]$. The area of the rectangle R_n with vertices *(n,0), (n + 1,0), (n,1/n), (n + 1,1/n)* is $1/n$, and the rectangle R_n contains the region H_n. If we let γ_n be the area of the region inside R_n but outside of H_n, then the series $\gamma_1 + \gamma_2 + \gamma_3 + \gamma_4 + \gamma_5 + \ldots$ may be shown to converge by the limit comparison test with $1/n^{3/2}$, and the sum is called γ, the *Eulerian gamma constant*. We then have, for large n, the following approximation for the partial sum of the harmonic series:

$$1 + 1/2 + 1/3 + \ldots + 1/n \approx ln(n + 1) + \gamma.$$

The small case Greek *gamma* was used by Euler (1707–1783) to denote the limiting value of $(1 + 1/2 + 1/3 + \ldots + 1/n) - ln(n + 1)$. It is unknown whether Euler's constant is rational or irrational. I have seen the symbol C used for this constant, for example, by Konrad Knopp. The value of γ is .57721566 to eight decimal places; there is a mnemonic device to remember this, but it requires knowledge of papal history. The year 772 was the year of the election of Adrian I, and 1566 was the year of the accession of Pius V to the supreme pontificate. Englishmen and Americans must learn to pronounce the name Euler *Oi-ler*, not *You-ler*.

gamma This is the third letter of the Greek alphabet, used in mathematics to denote a function of Euler and a probability distribution. In the mid-thirteenth century, the diagrams in mathematical manuscripts started to be labeled A, B, C,... in accordance with the Latin order of the letters, instead of A, B, G,..., the Greek order, as previously. The capital *gamma* Γ is the symbol for the gamma function, which generalizes the factorial function to the set of positive real numbers.

gematria This is the Hebrew גמטריא, the transliteration of, and therefore the Hebrew word for, *geometry*. It eventually acquired the specialized meaning of *the superstitious use of mathematics*, especially the deriving of meaning from the numerical value of the letters that compose words. The most famous case of this is Revelation XIII 18:

> Here is wisdom. Let him that hath understanding count the number of the beast: for it is the number of a man; and his number is Six hundred threescore and six..

The earliest occurrence of the word is in the Talmud (*Pirke Aboth* III, 23):

> Rabbi Eleazar Hisma said: Offerings of birds and purifications of women, these, yea these, are the essential precepts. Astronomy and geometry are but fringes to wisdom. (R. Travers Herford, *The Ethics of the Talmud: Sayings of the Fathers*, Schocken Books, New York, 1966, p. 93)

Actually, the word in this Talmudic text is in the plural number גמטריאות and so cannot mean *geometry*, as Herford translates, but must instead mean *mathematical calculations*.

gender One of the articles mentioned on the cover of the January 2012 issue of the *Notices of the American Mathematical Society* is "Debunking Myths about Gender and Mathematics Performance." The noun *gender* comes from the Latin *genus*, which means *type, kind*. The *d* crept in because of mispronunciation in Old French. Nouns in Latin, Greek, Russian, and German are divided into three types or kinds, masculine, feminine, and neuter. Nouns in Arabic, Hebrew, French, and Italian are of two types, masculine and feminine. English nouns have no such classes. Thus, *gender* is a technical term in certain languages conveying the information as to which class a noun belongs. *Sex* pertained to living beings and identified them as male or female. The use of *gender* in place of *sex* would formerly have been marked wrong, but it has now become commonplace. The noun *sex* is ever more being restricted to mean physical sexual activity. The topic is a hornets' nest.

The occasional use of *humankind* as an alternative to *mankind* is sanctioned by Gibbon and is unobjectionable.

generalization The Latin adjective *generalis* means *belonging to the genus*, that is to say, to the specific class or kind in question. It acquired the meaning of *relating to all* as opposed to *specific, relating to the species*, because the *genus* was a larger class than the *species*. The addition of the Greek verbal suffix *-ize* to the Latin stem, and the further creation of a noun based on a fictitious verb *generalizo, generalizare*, mark the word as a late, in this case eighteenth-century, invention.

generator This is the Latin noun for *begetter*, from the first-conjugation verb *genero, to beget*.

Genoan Lottery The Genoan Lottery is described by Euler in his paper "Sur la probabilité des sequences dans la Loterie Génoise," published in 1767 on pages 191–230 of the *Proceedings* of the year 1765 of the Prussian Royal Academy of Sciences in Berlin; the contents are summarized by Isaac Todhunter on pages 245–247 of his *History of the Mathematical Theory of Probability*, Chelsea Publishing Company, New York, 1965, an unaltered reprint of the 1865 first edition published by Cambridge University Press. Suppose we print n tickets, numbered consecutively from 1 to n, and then randomly pick l of them without replacement. We win first prize if the tickets can be arranged into one sequence of l consecutive digits. We win second prize if the tickets can be arranged so that $l - 1$ of them are consecutive, but not all l. Euler calculated the probabilities of winning first prize and of winning second prize. The following more general problem is suggested by the lottery and was not answered by Euler: What is the probability that there are s_i sequences of length l_i, $i = 1, 2, \ldots, r$; $l_1 < l_2 < l_3 < \ldots < l_r$? Put $s = s_1 + s_2 + \ldots + s_r$, the total number of sequences. Fabrizio Polo proved the answer to be

$$_{n-l+1}C_s[s!/(s_1!s_2!\cdots s_r!)]/{_n}C_l,$$

where $_pC_q$ means p *choose* q. If X is the random variable whose value is the number of sequences in the list of n digits, then the mean of X is

$l(n - 1 + 1)/n$. Euler did not make the mistake, when designing lotteries, of arranging for the jackpot to exceed the cost of buying all the tickets. This happened in 1992 in the Commonwealth of Virginia. (See the article "Group Invests $5 Million To Corner Lottery Market" in the February 25, 1992, issue of *The New York Times*, pp. A1, A9.) To play in the state lottery, one paid $1 and chose a combination of six numbers from the first forty-four positive integers. Since there are 7,059,052 such combinations, a fellow could ensure victory by purchasing one ticket for each combination, at the cost of $7,059,052. The first prize alone was $27,007,364, and if one added the prize money for the second, third, and fourth prizes, which such a gambler would also win, the total jackpot was $27,918,561. In the eighteenth century, the city of Paris went bankrupt as a result of having to pay the winner of a poorly designed city lottery.

genre This is the French word for *kind*, equivalent to the Latin *genus*, which was used by Descartes, who defined the *genre* of a polynomial equation to be *n* if the degree was *2n – 1*.

genus The Latin noun *genus, generis* means *kind*. The plural of *genus* is *genera*. To write *genuses* is unprecedented and impossible.

geodesic From γῆ (contracted from γέα), *earth,* and δαίειν, *to divide,* comes the noun γεώδαισις. The addition of the adjectival suffix to the stem gives us γεωδαισικός, whence, upon removal of the nominative ending -ός, the English adjective is derived. (The Greek diphthong αι was regularly transliterated by *æ* when the word came into Latin, and then the *æ* became *e*. This is how, for example, αἵρεσις became *haeresis* and then *heresy*.)

geometric The Greek adjective γεωμετρικός, *pertaining to geometry,* was derived by the addition of the adjectival suffix -ικός to the stem of the noun γεωμετρία, *geometry*. The adjective *geometrical* is also used, a word formed late by tacking on the stem of the Latin adjectival suffix -*al* at the end of *geometric*; the choice which of the two to use in any specific instance is determined by custom. For example, one speaks of the *geometric distribution*, not of the *geometrical distribution*.

geometry This is the Greek γεωμετρία, which means *measurement of the earth*. It is the earliest branch of mathematics and arose, according to Herodotus, in Egypt at the time of the Pharaoh Senusret III (1878–1839 B.C.), whom the Greeks called Σέσωστρις. The annual inundation of the Nile required a science to preside over the surveying required to readjust the taxes.

...εἰ δὲ τινὸς τοῦ κλήρου ὁ ποταμός τι παρέλοιτο, ἐλθων ἂν πρὸς αὐτὸν ἐσήμαινε τὸ γεγενημένον· ὁ δὲ ἔπεμπε τοὺς ἐπισκεψομένους καὶ ἀναμετρήσοντας ὅσῳ ἐλλάσσων ὁ χῶρος γέγονε, ὅκως τοῦ λοιποῦ κατὰ λόγον τῆς τεταγμένης ἀποφορῆς τελέοι. δοκέει δέ μοι ἐντεῦτεν γεωμετρίη εὑρεθεῖσα ἐς τὴν Ἑλλάδα ἐπανελθεῖν.... (Herodotus, Book II, 109, 2–3)

...and if the river should take away anything from any man's portion, he would come to the king and declare that which had happened, and the king used to send men to examine and to find out by measurement how much less the piece of land had become, in order that for the future the man might pay less, in proportion to the rent appointed: and I think that thus the art of geometry was found out and afterwards came into Hellas also.... (G. C. Macaulay's translation)

The textbook of Euclid, which contained the elements of geometry, is one of the obvious candidates for the greatest book in the world. Pharaoh Sesostris is known in the history of art because of the beauty of his portraits, which are immediately recognizable because of his drooping eyelids. The statue of this pharaoh is the most important artifact in the Brooklyn Museum.

Gibbon Edward Gibbon (1737–1794), the chief personality of the Enlightenment, had the following to say about mathematics on page 427 of the fifty-second chapter (vol. 5, 1788) of the first edition of the *History of the Decline and Fall of the Roman Empire*: "The mathematics are distinguished by a peculiar privilege, that, in the course of ages, they may always advance and can never recede."

global From the Latin noun *globus, a ball, a sphere*, came the adjective *globalis, spherical.*

glottochronology This is the name of a subject that is the offspring of mathematics and linguistics. It tries to determine how long a language has been around. The word is formed correctly from the Greek γλῶττα (*language*), χρόνος (*time*), and λόγος (*reckoning*). There are two main theorems. The first is: If there is a list of N_0 "basic words" in a language, that is to say, words that are not technical terms or borrowed from another language, then, *t* millennia later, the number of words still in use in the language is given by

$$N(t) = N_0 e^{-kt},$$

where *k* is a universal constant equal to −0.217. The second is: If two languages descended from a common parent share *M* words in a list of N_0 "basic words," then the time *T* that they have led separate existences as different languages is given by

$$T = -2.30 \ ln(M/N_0).$$

Among the conclusions that this theory has reached is that the original homeland of the Athabaskan family of American Indian languages (which includes Chippewa, Chiricahua Apache, and Navaho) is Lake Athabaska in northern Canada.

gnomon The Greek verb γιγνώσκω means *to know*. From it is derived the noun γνώμων, *one who knows, a judge, a carpenter's rule or square*. That portion of a parallelogram shaped like a carpenter's rule was therefore called a γνώμων by Euclid.

-gon This suffix is derived from the Greek noun γωνία, which means *a corner, an angle, a carpenter's square*. It is the second ingredient of several words describing plane geometric figures. Thus, the adjectives πολύγωνος, τρίγωνος, τετράγωνος, and πεντάγωνος mean, respectively, *having many angles, having three angles, having four angles*, and *having five angles*. The English words are produced by removing the masculine nominative case ending, except in the case of τρίγωνος; we do not use *trigon*, preferring instead the word *triangle* taken from Latin.

goniometry This is the name for all the theorems that can be deduced from the prostaphairesis and periodicity formulas of the trigonometric functions. It is composed of the Greek words γωνία, *an angle*, and μέτρον, *measure*, and means *the measurement of angles*. It is nowadays uncommon and should be avoided by those eager to be understood.

grad The Latin noun *gradus, gradūs, step*, comes from the verb *gradior, gradi, gressus, to walk*. The authors of the metric system removed the nominative case ending *-us* to produce the noun *grad*. It is the hundredth part of a right angle, the centesimal method of measurement being preferred by those authorities to the primitive Babylonian division of the right angle into ninety parts. It makes some sense because by dividing the circle into four hundred equal parts, one can determine the quadrant of a given angle from the first digit of the angle's measure. The grad has prevailed neither over the radian nor over the degree.

grade This is the Latin fourth-declension noun *gradus* made into an English word.

grade point average You cannot legitimately take the mean of an ordinal variable. However, this is often done in the social sciences. The *grade point average* is a weak reed for a school to rely upon as a measure of academic achievement, for when one leans upon it, it snaps.

gradient This word comes from the present participle *gradiens gradientis* of the Latin verb *gradior, to walk*. The spelling *gredient* is wrong. When the Latin verb *gradior* is compounded with prefixes, it becomes *-gredior*, as in *progredior*; the form *gredior* does not stand by itself.

gram This noun is derived from the Greek nouns τὸ γράμμα, *a letter*, and ἡ γραμμή, *a stroke in writing, a line*. It is an invention of the French National Assembly.

graph From the Greek verb γράφω, *to write or draw*, came the noun γραφή, *a drawing*.

gravitation See the entry **gravity** below. From the verb *gravo, gravare* there was formed the late frequentative verb *gravito, gravitare, gravitavi, gravitatus* with the meaning *to weigh down constantly*, and from its fourth principal part was formed the noun *gravitatio, gravitationis* in the usual manner.

gravity The Latin adjective *gravis, grave* means *heavy*, and is related to the Greek adjective βαρύς with the same meaning. From the adjective there was formed the verb *gravo, gravare, gravavi, gravatus* with the meaning *to load, weigh down*, and from this verb was formed the noun *gravitas, heaviness*, whence came the English noun.

gyration The Greek noun γῦρος means *a ring or circle*. The Latin ending *-atio, -ationis* was then added to the stem to form this macaronic noun. It is therefore a low word.

H

harmonic The Greek verb ἁρμόζω means *to fit together, to join*, especially of carpenter's work. From it was formed the noun ἁρμονία meaning *fitting together, joining, agreement, musical concord*. The addition of the adjectival suffix -ικός produced the adjective ἁρμονικός with the meaning *skilled in music*; the Greeks used this word in its mathematical sense, for example, they spoke of the *harmonic mean*. If O is an open subset of R^n, a real valued function f

defined on O is *harmonic* if $\partial^2 f / \partial x_1^2 + \ldots + \partial^2 f / \partial x_n^2 = 0$ everywhere on O.

haversine This is an absurd word of the nineteenth century meaning *half the versed sine*, that is, *(1 − cos θ)/2*. The formation of words in this way is buffoonery. A similarly contemptible word is *coversine*, which means the *complemental versed sine*, and is supposed to mean *1 − sin θ*; the authors of it were clumsy butchers and did not bother to take the complete first syllable of the first word. The associated abbreviations *hav θ* and *covers θ* are ludicrous.

-hedron The Greek verb ἕζομαι means *to sit*, and from it is derived the noun ἕδρα, which means *seat, base*. The adjective πολύεδρος, πολύεδρον means *having many faces*, τετράεδρος, τετράεδρον means *having four faces*, ἑξάεδρος, ἑξάεδρον means *having six faces*, ὀκτάεδρος, ὀκτάεδρον means *having eight faces*, δωδεκάεδρος, δωδεκάεδρον means *having twelve faces*, and εἰκοσάεδρος, εἰκοσάεδρον means *having twenty faces*. The second or neuter form of each adjective was then used alone with the noun σχῆμα, *figure*, understood, and we obtained the technical terms *polyhedron, tetrahedron, hexahedron, octahedron, dodecahedron,* and *icosahedron.*

helicoid This is a modern word compounded of ἕλιξ, ἕλικος, *a coil*, and the syllable *-oid* from εἶδος, *shape*. The word is used by Konrad Knopp in his discussion of the Riemann surface of the function $w = z^{1/2}$ (*Theory of Functions, Part II*, Dover Publications, Inc., New York, 1947, p. 102).

helix This word is the Greek ἕλιξ, which means *a coil*. It comes from the verb ἑλίττω, *to turn around*. The Greek plural, which is mandatory, is *helices*. To write *helixes* is very low since it has never been used by the best authors. Helical columns were employed in the ancient Middle East, as one can see on the cover of the February 1984 issue of the *Yale Alumni Magazine and Journal*. The most famous example of these are those of the baldacchino of St. Peter's Basilica by Bernini; he used six such columns at the Val de Grâce in Paris. Sir Christopher Wren intended the same for St. Paul's in London, and his plan was consummated after the Second World War. Double helical columns are to be found in the cloisters of St. Paul's outside the Walls in Rome. The Columns of Trajan and Marcus Aurelius in Rome are inscribed with engravings climbing up as a helix. Columns engraved with helices may be found in Durham Cathedral. The Canopy over the High Altar of St. Mary Major's in Rome is supported by four columns on which strands of laurel climb as a helix. Bramante constructed a helical staircase in the Apostolic Palace of the Vatican. In the following century, Bernini constructed a similar staircase in the Palazzo Barberini; St. Paul's Cathedral in London has a conical elliptical helical staircase. A similar conical helical staircase can be seen at the Supreme Court in Washington. At the entrance to the Vatican Museums, tourists exiting by means of the modern conical circular helical staircase automatically throw money into the huge vase at the bottom that they have been circling as they descend.

hemicontinuous This is an incorrectly formed synonymn of *semi-continuous (q.v.)*, the marriage of the Greek ἡμι- (*half*) and the Latin *continuus*. There is no need to use a Greek prefix before a word of Latin origin when there is a Latin prefix available.

hemisphere The Greek adjective ἥμισυς means *half*; the first three letters ἡμι- were used as a prefix like the English *half-*. Thus, ἡμίσφαιρος means *half a sphere*. The Romans used the prefix *semi-* for this purpose. Their word *semis, semissis* meaning *the half of anything*, is etymologically related to ἥμισυς.

heptagon The Greek word ἑπτά means *seven*. The combination of this numeral and the noun γωνία meaning *angle* produces the word ἑπτάγωνον, *heptagon*, actually the neuter singular of the adjective ἑπτάγωνος meaning *having seven angles*. See the entry **-gon**.

hereditary The Latin noun *heres, heredis* means *heir*, the noun *hereditas, hereditatis* means *inheritance*, and the adjective *hereditarius, -a, -um* means *pertaining to an inheritance*.

heterodyne The *heterodyne theorem*, also called the *modulation theorem*, is discussed by Nahin in *Dr. Euler's Fabulous Formula*, Princeton University Press, Princeton and Oxford, 2006. The word is not formed correctly, as there is no word *dyne* in Greek. The Greek word for *power* is δύναμις, and the adjective meaning *powerful* is δυνατός. The Greek adjective ἕτερος means *other*. The correct adjective would have been *heterodynate*.

heterological The Greek adjective ἕτερος means *other*, and the noun λόγος means *word, reason*. The word *heterological* was thrown together with both Greek and Latin parts. *Heterologic* would have been sufficient as an English adjective since *-ic* is the remnant of the Greek adjectival suffix -ικός, but that fact was not recognized by those who superimposed the Latin adjectival suffix *-al* (minus the case ending) to produce *heterological*. The word is central to the Grelling-Nelson paradox: An adjective is *heterological* if and only if it does not apply to itself. Thus, *long* is heterological. Is heterological itself heterological?

heuristic This is a word of the nineteenth century. The Greek verb εὑρίσκω means *to find*. Someone imagined that the irregular formation εὑριστικός would mean *helping to find*, but there is no such Greek word.

hexagon See the entry **-hedron**.

hexagram The Greek word ἕξ means *six*, and γράμμα means *line*. The adjective ἑξάγραμμος, ἑξάγραμμον therefore means *having six lines*. That the English word has one instead of two *m*'s is a result of the interference of Noah Webster, who canceled (Johnson would write *cancelled*) the consonant because the doubling is not noticed in English pronunciation.

hippopede The Greek noun πέδη means *fetter*, and ἵππος means *horse*; the compound ἱπποπέδη, therefore, means a *horse-fetter*. It is the name of a plane curve attributed to Proclus, one of the later (fifth century A.D.) successors of Plato as head of the Academy and the commentator on the first book of Euclid's *Elements*. The polar equation is $r^2 = 4b(a - b \sin^2 \theta)$, where a and b are positive constants. The experienced reader will notice that if $b = 2a$, the hippopede is a lemniscate. See Lawrence, pages 145–146.

histogram The Greek noun ἱστός means the *mast of a ship*; it is derived from the verb ἵστημι, *to stand*. *Gram* is from the Greek γράμμα, *anything drawn, a letter, a picture*, from γράφω, *to grave, sketch, write*. It is a type of bar graph common in probability and statistics, which displays the number of times a random variable takes values in one of many subintervals of equal length into which its range is divided.

holomorphic The Greek adjective ὅλος means *whole, entire, complete.* From it both the Greeks and the English derived the prefix *holo-.* *Morphic* is a modern invention derived by attaching the stem of the Greek adjectival suffix -ικός to the stem of the noun μορφή, *a form, shape, or figure.* A holomorphic function is a complex analytic function.

homeo- This prefix is derived from the Greek adjective ὅμοιος, which means *like, resembling.*

homeomorphic See the entry **homeomorphism**. Two topological spaces are *homeomorphic* is there is a homeomorphism between them. The *-ic* is the stem of the Greek adjectival suffix -ικός.

homeomorphism See the entries **homeo-** and **morphism**. This word is an invention of the modern topologists. A *homeomorphism* is a continuous function from one topological space to another that has a continuous inverse.

homo- This prefix is derived from the Greek adjective ὁμός, which means *one and the same.* The corresponding Latin prefix is *idem-.* The debate of whether ὁμοούσιον or ὁμοιούσιον was the right word to put into the Nicene Creed led to religious war in the fourth century. In mathematics, however, the difference between *homeomorphic* and *homomorphic* is mainly that the topologists prefer the former word and the algebraists the latter.

homogeneity This modern noun has an ending derived from Latin (-*ty* is the transfiguration of -*tas*) appended to a noun derived from Greek (ὁμογένεια). See the following entry.

homogeneous This word is derived from the medieval Latin adjective *homogeneus, -a, -um* formed from the Greek noun ὁμογένεια, which means *having the same* (ὁμός) *origin* (γένεια). *Homogeneity* is the property of a polynomial whose terms are all of the same degree, where in determining the degree, one counts the powers of the coefficients as well as the powers of the variable. For example, the formula $ax^2 + b^2x + c^3$ is *homogeneous* because the degrees of the three terms are all 3. From the modern point of view, one may write $ax^2 + bx + c$ instead of $ax^2 + b^2x + c^3$, which latter expression was required by the traditional geometrical interpretation of the polynomial as the sums of various cubes. A differential equation is *homogeneous* if it can be written in the form $y' = f(y/x)$; for example, the equation $y' = (x + y)/(x - y)$ is homogeneous since $y' = [1 + (y/x)]/[1 - (y/x)]$. Such equations become variables-separable equations by the substitution $v = y/x$. The equation of the example arises in the problem of determining those curves that have the property that the slope at any point is equal to the sum of the coordinates divided by their difference; the solutions are logarithmic spirals from which the points where $y = x$ have been removed.

homologous See the entry **homo-** above. The suffix -*logous* is made from the noun λόγος, *word, reason, ratio*. The Greek adjective ὁμόλογος means *agreeing* and is compounded of the prefix ὁμο- and the noun λόγος. If $a/b = c/d$, then a and c are *homologous in the ratio*, and b and d are *homologous in the ratio*.

homology See the entry **homo-** above. The suffix -*logy* is made from the Greek noun λογία, a collection, from λέγω, *to gather*. The appropriate corresponding adjective is *homologic*.

homomorphism See the entries **homo-** above and **morphism**. Herstein (*Topics in Algebra,* Blaisdell Publishing Company, 1964, p. 46) defines a *homomorphism* to be a function φ from one group (G₁, •₁) to another (G₂, •₂) that preserves the structure, that is, such that

$$\varphi(a \bullet_1 b) = \varphi(a) \bullet_2 \varphi(b).$$

If the homomorphism is one-to-one, he calls it an *isomorphism*. (Others call this a *monomorphism*.) An *automorphism* he defines to be an isomorphism of a group onto itself. If the isomorphism is onto the group (G₂, •₂), he says that the two groups are *isomorphic*. He does not use the term *homomorphic*.

homoscedasticity If $X_1, X_2, \ldots$ is a sequence of random variables that all have the same finite variance, then the random variables are called *homoscedastic*. *Homoscedasticity* is the property of being homoscedastic. The Greek verb σκεδάννυμι means *to scatter*, the derived noun σκέδασις means *scattering*, the adjective σκεδαστικός means *able to disperse*, and the adjective σκεδαστός means *scatterable*. The English *homoscedastic* is the stem of σκεδαστικός. Someone attached the nominal ending *-ity* to the stem of one of the two adjectives to produce the macaronic term *homoscedasticity*. It is macaronic because the suffix *-ity* is the sign of the French *-ité*, which proceeds from the Latin *-itas*.

homothetic See the entry **homo-** above. The word *homothetic* is a modern invention created on the analogy of adjectives like *synthetic* from the verb τίθημι, *to put or place*. The associated adjective is θετικός, which means *fit for placing, apposite, positive*. A homothetic function of the plane is a dilation or contraction followed by a translation, that is, a function that takes *(x,y)* to *(u,v)* where $u = h + \lambda x$ and $v = k + \lambda y$, where h, k, and λ are arbitrary real numbers. The idea intended is that geometric figures are placed in the same position relative to one another after the application of the homothetic function as they were before. However, the adjective θετικός never meant *placed*, which was τιθείς.

homotopic See the entry **homo-** above. *Topic* is the stem of the Greek adjective τοπικός, *pertaining to position*, from the noun τόπος, which means *place*. The same noun is the origin of *-topy* in *homotopy*. *Homotopic* and *homotopy* are modern words.

homotopy See the entry **homotopic** above.

horizon, horizontal The Greek noun ὅρος means *boundary*. From it is derived the verb of action ὁρίζω, *to draw the boundary*. The verb's present active participle is ὁρίζων, ὁρίζοντος, which was used as an adjective modifying the noun κύκλος, which means *circle*. The horizon was therefore *the bounding circle*. The adjective was formed late as if the word were of Latin origin; it would more correctly have been *horizontic*.

hour This is the Latin *hora*.

hyper- This prefix is the Greek preposition ὑπέρ, which means *beyond*. The corresponding Latin prepositions are *trans* and *ultra*, and the corresponding English prefix is *over-*.

hyperbola The *hyperbola* is the set of all points in the plane, the difference of whose distances from two fixed points is fixed. The constant difference of the distances is usually denoted by *2a*, and the distance between the two fixed points is usually denoted by *2c*. The parameter *a* must be less than the parameter *c*, so we define a third parameter *b* by $b^2 = c^2 - a^2$. The ratio *2c/2a* is called the eccentricity *e* of the hyperbola. We must have *1 < e*. The *latus rectum* is $2b^2/a$. The word *hyperbola* is the Latinization of the Greek ὑπερβολή, which means *excess*, from the verb ὑπερβάλλω, which means *to fall beyond*. The origin of the name is as follows. Consider the hyperbola whose equation is $[(x + a)^2/a^2] - y^2/b^2 = 1$. Let *P(x,y)* be a point on the hyperbola not a vertex, and let *S* be a square of side $|y|$. Let *R* be a rectangle whose base is *x* and whose altitude is the length of the latus rectum of the hyperbola. Then the area of *S* is greater than (that is, exceeds) the area of R.. The hyperbola may also be defined by the focus-directrix definition: Let *e > 1*, let ℓ be a fixed line (the *directrix*) and *F* a fixed point (the *focus*) a distance *d* (the *directral distance*) from ℓ, *d > 0*. Then the hyperbola is the locus of all points *P* such that

$FP/F\ell = e$. If F is the pole and ℓ is the line with equation $x = d$, then the polar equation of the hyperbola is $r = ed/(1 + e\cos\theta)$. The directral distance is related to the other parameters by the formula

$$d = a(e^2 - 1)/e.$$

hyperbolic The addition of the adjectival suffix -ικός to the stem of the noun ὑπερβολή produced the adjective ὑπερβολικός with the meaning *pertaining to the hyperbola*. The **hyperbolic functions** were the invention of the Alsatian mathematician Johann Heinrich Lambert (1728–1777), born in Mülhausen on August 26, 1728. His most famous achievement was the proof that π is irrational. Frederick the Great admitted him to membership in the Berlin Academy of Sciences. When the king inquired in what branch of science Lambert was most proficient, he replied, "In all of them." If one allows a radius vector from the origin to rotate counterclockwise from the positive x-axis so as to sweep out a plane region with area equal to $\theta/2$ between itself, the x-axis, and the "unit hyperbola" with equation $x^2 - y^2 = 1$, then one calls the coordinates of the point of intersection of the vector with that hyperbola *(cosh θ, sinh θ)*. One then defines *tanh θ, coth θ, sech θ,* and *csch θ* by analogy with the trigonometric definitions. The familiar formulas for *cosh θ* and *sinh θ* in terms of the exponential function follow from considering the area of the triangle with vertices the origin, *(cosh θ, sinh θ)*, and *(cosh θ, 0)*. What we call the *natural logarithim* was called by Euler the *hyperbolic logarithm*.

hyperbolic spiral The hyperbolic spiral, first studied by Varignon (1654–1722) in 1704, is the polar curve with equation $r = 1/\theta$, it is called *hyperbolic* because of the similarity of its equation to $y = 1/x$. Two hyperbolic spirals may be discerned on the eighteenth-century façade of the Church of Santa Maria in Campitelli in Rome, the work of Carlo Rinaldi.

hyperboloid This word is a modern invention made on the analogy of the Greek words *rhomboid* and *trapezoid*, the result of adding the suffix *-oid* to the stem of the noun *hyperbola*. It is formed from the combination of the two words ὑπερβολή, *hyperbola*, and εἶδος, *shape*;

it should mean *something that looks like a hyperbola*. However, since the hyperboloid, being solid, does not look like a hyperbola, which is a plane figure, the word does not convey the meaning it was intended to. Observe that the real Greek words *rhomboid* and *trapezoid* referred to plane figures, not solids.

hypercomplex This modern word is macaronic, the compound of the Greek preposition ὑπέρ and the stem of the Latin adjective *complexus*. A better name would have been *ultracomplex*. See the entry **complex**. According to Herstein (*Topics in Algebra*, p. 218), an associative ring A is called an *algebra* over a field F if A is a vector space over F such that for all $a, b \in A$ and $\alpha \in F$, $\alpha(ab) = (\alpha a)b = a(\alpha b)$. If the field F is the field of real numbers or the field of complex numbers, the elements of A are called *hypercomplex numbers*.

hypercube See the entries **hyper-** and **cube**. The intention of the authors of this word is that the prefix *hyper-* imply that the cube is being generalized. An n-dimensional hypercube is any rigid motion of the convex hull of the points $(x_1, x_2, x_3, \ldots, x_n)$, where each x_i is either 0 or 1.

hypergeometric See the entries **hyper-** and **geometric**. The hypergeometric distribution is the generalization of the geometric distribution of probabilities. The force of the prefix *hyper-* in modern concoctions is usually to indicate a generalization.

hyperplane This is a low word, the union of the Greek preposition ὑπέρ and the Latin noun *planus*. *Ultraplane* would have been better. The word is intended to generalize to n-dimensions the line of two dimensions and the plane of three dimensions. It may be defined as the solution set of the equation $a_1 x_1 + \cdots + a_n x_n = 1$, where the coefficients are not all zero.

hyperspace This is a low word, the union of the Greek preposition ὑπέρ and the Latin noun *spatium*. *Ultraspace* would have been better. The original meaning was *any Euclidean space of dimension greater than three*.

hypo- This prefix is the Greek preposition ὑπό, which means *under*. The corresponding Latin prefix is *sub-*. The corresponding English prefix is *under-*.

hypocycloid This correctly formed word is the invention of the astronomer Römer (1674). If a circle of radius b rolls without slipping inside a fixed circle of radius a, $a \geq b$, the trajectory of a fixed point P on the rolling circle is a *hypocycloid*. For the etymologies, see the entries **hypo-** and **cycloid.** In Euler's Latin, the curve is *hypocyclois, hypocycloidis*. The parametric equations of the hypocycloid are

$$x = (a - b)\cos \theta + b \cos [(a - b)/b]\theta$$

$$y = (a - b)\sin \theta - b \sin[(a - b)/b]\theta.$$

If $a = b$, the locus is a point. If $a = 2b$, the hypocycloid is a diameter of the stationary circle. If $a = nb$, $n = 3, 4,...$, the hypocycloid has exactly n cusps; if a and b are incommensurable, there are infinitely many cusps. When $n = 3$, the hypocycloid is called a *deltoid*; if $n = 4$, it is called an *astroid*.

    Newton proved the following theorem: Let A and B be two points on the earth's surface. A tunnel is dug through the earth from A to B, and a point mass travels from one point to the other entirely under the force of gravity. Then the time required for the trip is least when the tunnel is the arc of a hypocycloid connecting A to B with cusps at A and B but no intermediate cusp. See the entry **tunnel.**

    Euler proved the following two theorems, which were published posthumously (*De duplici genesi tam Epicycloidum quam Hypocycloidum*, *Acta Petropolitana*, 1784, pp. 48–59, §17 and §4):

1) If a_1, a_2, and a are three positive numbers such that $a_1 + a_2 = a$, then the hypocycloid produced by rolling a circle of radius a_1 inside a circle of radius a is congruent to the hypocycloid produced by rolling a circle of radius a_2 inside the circle of radius a.

Talis circulus, cuius radius $b = (a + c)/2$, eandem Hypocycloidem describet ac minor circulus, cuius radius $b = (a - c)/2$.... Omnes Hypocloides duplici modo generari posse, quandoquidem eadem curva describitur, sive radius circuli mobilis sit $(a - c)/2$ sive $(a + c)/2$, quemadmodum deinceps sum demonstraturus.

2) If we allow $a < b$, then the hypocycloid produced when the radius b of the rolling circle exceeds the radius a of the stationary circle is congruent to the epicycloid produced by rolling a circle of radius $b - a$ about a circle of radius a.

Augeamus nunc alterius circulum mobilem B, ut circulum fixum A superet eumque totum in se complectatur, ita ut sit $b > a$; tum autem si punctum contactus initio sit in C, ubi simul stilus concipiatur, provolutione huius circuli B circa fixum A curva describatur CZ, tota extra circulum fixum sita, quae ergo iterum ad classem Epicycloidum erit referenda, atque adeo eadem erit, quae prodiret, si circulus mobilis, cuius diameter foret = DE, excessui scilicet diametrorum CD super CE aequalis, sive cuius radius foret = b-a, extra circulum fixum, qualis in figura est circulus Cd, revolveretur...

hypotenuse This word is derived from the Greek feminine present participle ὑποτείνουσα of the verb ὑποτείνω, which means *to stretch* (τείνω) *under* (ὑπό). The adjective modified the noun γραμμή, *line*, understood. The spelling *hypothenuse* with the *h* is wrong since the Greek letter is *tau*, not *theta*.

hypothesis An ὑπόθεσις in Greek is *that which stands* (θέσις) *underneath* (ὑπό) *something else.*

hypotrochoid If a circle of radius b rolls without slipping inside a fixed circle of radius a, $a \geq b$, and if a fixed point P is at a distance c from the center of the rolling circle, then the locus of P as it rolls along with the circle is called a *hypotrochoid*. For the etymologies, see the entries **hypo-** and **trochoid**. The hypotrochoid is called either *curtate* (from the Latin *curtatus*, which means *shortened, reduced*) or *prolate* (from the Latin *prolatus*, which means *brought forward, extended*).

according to whether $c < b$ or $c > b$. The parametric equations of the hypotrochoid are

$$x = (a - b)cos\ \boldsymbol{\theta} + c\ cos\ [(a - b)/b]\boldsymbol{\theta}$$

$$y = (a - b)sin\ \boldsymbol{\theta} - c\ sin[(a - b)/b]\boldsymbol{\theta}.$$

I

-ical Words ending in this way are usually the offspring of ignorance. The stem *-al* of the Latin adjectival suffix *-alis* has been superimposed upon an adjective ending in the stem *-ic* of the Greek adjectival suffix -ικός. This was done because those who did so did not recognize that the word they were dealing with was already an adjective; it was all Greek to them. For example, from the noun μάθημα, μαθήματος, *learning,* the Greeks formed their adjective μαθηματικός, *pertaining to learning.* The Romans adopted this adjective, which in their language became *mathematicus.* There is no Latin word *mathematicalis* since *mathematicus,* being already an adjective, did not require the addition of the suffix *-alis* to its root to make it so. The adjective *mathematicus* became in English *mathematic* or *mathematick,* and is a recognized adjective in *Johnson's Dictionary.* Some authors, however, superimposed the Latin ending on the word *mathematic* to produce *mathematical,* and the latter adjective, which also appears in *Johnson's Dictionary,* has displaced the former, which has altogether disappeared from modern prose. In some instances, however, as in the case of *dynamic* and *dynamical,* both the true and the inflated adjectives have survived. Thus, we hear both of the method of translation called *dynamic equivalence* and of the subject *dynamical systems.* As a general rule, when making new words, *-al* should not be added to what are already adjectives of Greek origin ending in *-ic.* In the case of words like *mathematical* and *dynamical,* their use is sanctioned by immemorial custom.

170

icosian This is a word invented by William Rowan Hamilton for the name of one of his games. The Greek word for *twenty* is εἴκοσι; to this word Hamilton added the stem of the Latin adjectival ending *-anus* to produce *icosian*. The game consists of starting at one vertex of a dodecahedron and moving along the edges until one has touched every vertex exactly once but has passed through no edge more than once; the path must be a cycle, that is, it must end at the vertex whence it began. The solution path has twenty edges, hence the name of the game.

ideal The Greek noun ἰδέα means *something seen by the eye of the mind*, from ἰδεῖν, *to see*. It was taken over into Latin as *idea, ideae*, from which the adjective *idealis* meaning *existing as an idea* proceeded.

idempotent The Romans did not use the adjective *idem*, *same*, as a prefix; therefore, whoever coined this word made a mistake. The participle *potens, potentis* of *possum, to be able*, means, *capable, powerful.*

identity From the Latin adjective *idem*, which means *the same*, there arose in the fifth century the late Latin noun *identitas, identitatis* with the meaning *sameness*.

-ικός, -ικη, -ικόν The Greeks added this suffix to the stem of a noun to create an adjective denoting relation, fitness, or ability. Hence when such words came into Latin, the ending was modified to *-icus, -ica, -icum*. The ending *-ic* on an English adjective is a give-away that it has this origin. When forming adjectives, it is incorrect to superimpose this suffix on a root that is not Greek. For example, with a Latin root one uses the corresponding suffix *-alis, -ale*.

image The Latin word *imago, imaginis* means *a likeness*. It entered French and then English as *image*. Words of Latin origin that are the same in French and English entered the English language as a result of the Norman invasion of 1066.

imaginary The Latin adjective *imaginarius* was formed by adding the adjectival suffix *-arius* to the stem of the noun *imago, imaginis*, which

171

means *likeness*. The English adjective is the Latin word without the case ending *-us*, the *i* having been changed, *per bellezza* as the Italians say, to *y*. The dreary names *imaginary* and *complex* were applied to the set of numbers of the form $a + b(-1)^{1/2}$ which, if $b \neq 0$, had no place on the number line; such entities, it was thought, were like phantoms and quite beyond comprehension.

implication The Latin noun *implicatio, implicationis* means *an entangling, an entanglement*. It is formed from the fourth principal part of the verb *implico*. See the following entry.

implicit The Latin verb *implico, implicare, implicavi, implicatus* means *to involve, to fold (plico) in (in)*. There was a rare adverb *implicite* meaning *intricately*, and a frequentative verb *implicito, implicitare, implicitavi, implicitatus* meaning *to keep on entangling*.

imply This verb is derived from the Latin verb *implico, implicare, implicavi, implicatus*, which means *to involve, to fold (plico) in (in)*. The mathematical meaning must have arisen from the idea that when we say that *A* implies *B*, we involve *B* as a consequence of *A*.

improper The prefix *in-* negates Latin adjectives. The Latin adjective *proprius* means *fitting, appropriate*. Thus, *improprius* means *not befitting*.

in- The Latin prefix *in-* may be either the preposition *in*, which means the same as the English word, or the negating prefix meaning *not*.

incenter See the entry **incircle** below.

inch This is the corruption of the Latin *uncia*, whence we also get *ounce*. It is a general unit of small measurement whether of length or weight.

incident The Latin verb *incĭdo, incĭdere, incĭdi, incasus* means *to fall (cado) on (in)*. Its present participle is *incidens, incidentis*, from whose stem the English adjective is derived. The Latin verb is to be distinguished from *incīdo, incīdere, incīdi, incīsus*, which means *to cut (caedo) into*.

incircle This is a low, confusing word. It is the circle inscribed in a triangle, and its center is called by an equally low and confusing word, the *incenter*.

inclination The Latin verb *inclino, inclinare, inclinavi, inclinatus* means *to bend*. It is related to the Greek verb κλίνω, *to bend*. Unlike the Greeks, the Romans never used the form *clino* by itself. The noun is the stem of the Latin noun *inclinatio, inclinationis*, which means *leaning, bending*.

include The Latin verb *claudo, claudere, clausi, clausus* means *to close*. When compounded with the preposition *in*, it produces the verb *includo, includere, inclusi, inclusus* with the meaning *to close in*.

incommensurable The modern Latin word *incommensurabilis* is well constructed and would mean *not able to be measured together with something else*. The prefix *in-* indicates *not*, the prefix *com-* indicates *together with*, the noun *mensura* is *measure*, and the suffix *-abilis* indicates ability.

incompatible The medieval ecclesiastical Latin word *compatibilis* was applied to church livings and meant *permitted to be held together with*. Thus the archdiocese of York and the abbey of St. Albans were compatible benefices and in fact once held by the same person, Cardinal Wolsey. *Compatible* is derived from the prefix *com-* from *cum, together*, and the verb *patior, pati, passus, to suffer, to permit*. The addition of the negating prefix *in-* results in the word *incompatibilis* with the meaning *not suffered to be held together with*.

incomplete This word comes from the Latin adjective *incompletus* of the same meaning. The prefix *in-* here is the Latin negating prefix, so the word means *not complete*. See the entry **complete**.

inconsistent The prefix *in-* here is the negating prefix, so the word means *not consistent*. It is a modern word. See the entry **consistent**.

increase This word is derived through the mediation of French from the Latin *incresco*. See the following entry.

increment This word is the root of the Latin noun *incrementum*, meaning *growth* or *increase*, which is derived from the verb *incresco, increscere, increvi*, which means *to grow* (*cresco*) *in* or *upon* anything. Latin "nouns denoting *acts* or *means* or *results* of acts are formed from roots or verb-stems by the use of the suffixes *-men, -mentum, -monium*, and *-monia*" (Allen and Greenough, §239).

indefinite This word exists in Latin as the adverb corresponding to the adjective *indefinitus, -a, -um*, which means *indefinite*. The prefix *in-* here is the negating prefix. See the entry **definite**.

in-degree The *in-degree* of a directed graph at a vertex is the number of edges leading into the vertex. This is a sensible modern word, unlike *incenter* and *incircle*. See the entry **degree**.

independence The Latin word for what we call *independence* is *libertas*. The words *independent* and *independent* came into English around the year 1600 having first secured a place in the French tongue. See the entry **dependence**.

independent See the preceding entry.

indeterminate The Latin noun *terminus* means *the end of* something. From it was formed the denominative verb *termino, terminare, terminavi, terminatus* meaning *to set bounds to*. The addition of the prefix *de-* emphasizes that the separation is *from* something else and produced the compound verb *determino, determinare, determinavi, determinatus* meaning *to fix the limits of*. The addition of the negating prefix *in-* to the past participle of this verb resulted in the Latin adjective *indeterminatus* meaning *undefined, unlimited*.

index The Latin verb *dico, dicare, dicavi, dicatus* means *to consecrate, to dedicate, to make known as devoted*. The addition of the preposition *in* as a prefix produced the compound *indico, indicare, indicavi, indicatus* with the meaning *to make known*. The noun *index, indicis* means *the person or thing that does the informing*. The plural of this word is *indices*. The English plural *indexes* is permissible.

indicator See the preceding entry. The addition of the nominal suffix *-or* to the stem of the fourth principal part of the verb *indico, indicare* produces the noun of agent *indicator* meaning *the one who points out*.

indicatrix This word is the feminine of *indicator* and means *a female who points out*.

indirect This is an English word composed of Latin parts. The prefix *in-* here is the negating prefix. See the entry **direct**.

individual This word is the stem of the medieval Latin adjective *individualis, -e* used by the eleventh-century translator of the Arabic Euclid, Adelard of Bath, to indicate something considered as a unit for the matter at hand. In particular, he spoke of the *formae individuales*, the indivisible ideas. It was built upon the classical Latin *individuus, -a, -um* which translated the Greek ἄτομος, which means *undivided*. The word *individuus* is formed of the negating prefix *in-* added to the stem of the verb *divido, dividere, divisi, divisus, to divide*. The individual ergodic theorem of Birkhoff (1931) says that if T is a measure preserving transformation on [0,1] and if f is a measurable, real-valued function in L_1 with domain [0,1], then the limit of $[f(0) + f(T(x)) + f(T^2(x)) + \cdots + f(T^{n-1}(x))]/n$ as n approaches infinity exists almost always and defines a function f^* in L_1 that is T-invariant.

induce This verb is derived from the Latin *induco, inducere*. See the entry **induction**.

induction The Latin verb *induco, inducere, induxi, inductus* means *to lead (duco) in (in)*. The noun *inductio, inductionis* with the meaning *a leading or bringing in* is formed from the fourth principal part. It is used by Cicero to translate Aristotle's technical term ἐπαγωγή. *Induction* is the method of drawing conclusions from experimental facts rather than by reasoning according to the laws of mathematics from accepted postulates, which latter method is called *deduction*. The *weak principle of mathematical induction* says that if a subset M of the set N of natural numbers has two properties, then $M = N$. The two properties are 1) *1 ϵ M* (the induction step) and 2) $n \in M \Rightarrow n + 1 \in M$ (the hereditary step). If step 2 is replaced by

$$\{1, 2, 3,...,n\} \subseteq M \Rightarrow n + 1 \in M,$$

then the axiom is called the *strong principle of mathematical induction*.

inductive The Latin adjective *inductivus, -a, -um* means *relating to an assumption*. See the entry **induction**.

inequality The Latin noun *aequalitas* means *evenness, smoothness*. See the entry **equal**. The Latin noun *inaequalitas* means *unevenness, unlikeness, inequality*. The ending *-as* became the French *é*, which became the English *y*.

inequivalence See the entry **equivalence**. Once one has the medieval Latin noun *aequivalentia*, it is inevitable that someone should take the next step and admit *inaequivalentia*, but there is no evidence of the word. *Inequivalence* is a natural formation that I saw, however, for the first time in the entry "Inequivalence of two definitions of absolute continuity" in the index of Halmos's *Measure Theory* (p. 301). The word *Halmos's* affords the opportunity to discuss the Saxon possessive in nouns ending in sibilants. The criterion by which one decides whether it is recommendable to add *'s* to a word ending in *s* is the standard pronunciation of the word; which spelling corresponds to how the word is actually pronounced? For this reason *Descartes'* is preferable to *Descartes's*, though the latter is correct.

inertia Sylvester's *law of inertia* is a theorem on real symmetric matrices on page 310 of Herstein's *Topics in Algebra*. *Inertia* is a Latin noun meaning *want of skill*. The noun *ars* means *skill*, and the adjective *iners* means *lacking skill, slothful*.

inference This is the medieval Latin noun *inferentia* found, according to the *Oxford English Dictionary*, in Abelard. The Latin verb *infero, inferre, intuli, inlatus* means *to bring, bear, to carry (fero) in (in)*. The noun *inference* is formed from the present participle *inferens, inferentis* of this verb.

inferior This is the comparative degree of the Latin adjective *inferus*, which means *below*. It means *lower*.

infimum This is the superlative degree of the Latin adjective *infimus*, which means *low*. Thus *infimum* means *lowest*. The accent is on the first syllable; the pronunciation *in-fi´-mum* is wrong.

infinite The Latin adjective *infinitus* means *unbounded*. The prefix *in-* here is the negating prefix, which has been added to the past participle *finitus* of the verb *finire*. See the entry **finite**.

infinitesimal This is a very strange word. To produce this English word, the stem of the Latin adjectival suffix *-al* has been added to the stem of a modern Latin adjective *infinitesimus*. The Romans added the ending *-esimus* to cardinal numbers to make ordinal numbers; thus, *centum* is *a hundred*, and *centesimus* is *a hundredth*. On this analogy, the seventeenth-century mathematicians formed the word *infinitesimus* from *infinitus, infinite*. Thus, the *pars infinitesima* was $1/\infty$, which we know to be meaningless. The word *infinitesimal* was defined by Dr. Johnson to mean *infinitely divided*, which makes no sense at all.

infinitive The Latin adjectives *infinitus, -a, -um* and *infinitivus, -a, -um* mean *not (in-) bounded (finitus)*. From the latter proceeds the grammatical term *[modus] infinitvus* meaning *the infinitive [mood], the infinitive*. The headline on page B1 of the Business Day section of the December 5, 2011, edition of *The New York Times* reads "China Says It's Unable to Easily Aid Europe." What caught my eye here was the fact that *The New York Times* has stopped holding the line against the split infinitive. In the old days, the editor would have corrected *to easily aid Europe* to *to aid Europe easily*. I once read that only old fogeys complained about split infinitives, that the rule against them arose from the fact that in Latin the infinitive is one word and cannot be broken apart, but since in English the infinitive is two words, the sign of the infinitive *to* and the verb, there is no reason why we may not insert other words in between the *to* and the bare verb. This is very bad reasoning indeed. The most polished authors have managed to be polished without splitting infinitives. However,

> ...a real split infinitive, though not desirable in itself, is preferable to either of two things, to real ambiguity, and to patent artificiality. (Fowler, *Modern English Usage*, second edition, Oxford University Press, 1965, p. 581b)

This is just common sense, which is the Supreme Court in all matters of this sort.

infinity The prefix *in-* here is the negating prefix. Although Latin has no noun *finitas*, it does have the noun *infinitas* meaning *endlessness*. Struik says (p. 251) that the symbol ∞ was first used by Wallis (1616–1703), and, indeed, Cajori, in his index, calls it *Wallis's sign*. See the entry **finite**.

inflection The Latin verbs *flecto, flectere, flexi, flexus* and *inflecto, inflectere, inflexi, inflexus* both mean *to bend*. The noun *inflection* is a misspelling of the root of the noun *inflexio, inflexionis*. The phrase *inflection point* has enetered the vocabulary of the talking heads. On July 11, 2010, Dan Sidor appeared on the ABC Sunday morning program *This Week* and described as "an inflection point in policy" the change in the role of the United States in Afghanistan.

information The Latin noun *informatio, informationis* means *a conception, idea*. It is derived from the past participle of the verb *informo, informare, informavi, informatus,* which means *to impose a shape (forma) on (in)*.

initial The Latin verb *ineo, inire, inivi, initus* means *to enter on, to begin*. From its fourth principal part was produced the noun *initium*, which means *beginning*. The adjectival ending *-alis* was then added to the stem of *initium* to form the adjective *initialis*, whence is derived, by dropping the case ending *-is*, the English adjective *initial*.

injection The Latin verb *inicio, inicere, inieci, iniectus* means *to throw (iacio) into (in)*. From the fourth principal part of this verb was formed the noun *iniectio, iniectionis* meaning *a laying or throwing on*. The associated English adjective *injective* is modern.

inner This is the comparative degree of the Anglo-Saxon adjective *in*, which meant *interior*; it is cognate with the Latin preposition *in* and the Greek preposition ἐν.

innovation The Latin adjective *novus, -a, -um* means *new*, and from it is derived the verb *innovo, innovare, innovavi, innovatus* meaning *to renew, to alter*. From the fourth principal part of this verb comes the noun *innovatio* meaning *renewal, alteration, innovation*. Innovation may be positive, a productive new idea, or it may be negative, *novelty*, the commotion caused by destructive people eager to sweep away the accomplishments of the millennia and replace them with rubbish. Indeed, another Latin word for innovation, *res novae*, means *revolution*.

The annual reports that members of a Mathematics Department must submit to their deans always inquire whether the professor has been *innovative* in any way during the year in question. The word *innovation* has thus become cant. Among college administrators, the negative connotation of *innovation* is no longer recognized as in existence, however much it predominated in former times.

The innovations in mathematics education in the last half century have been remarkable. Here follow the eighteen questions that constituted the honors examination for the senior mathematics majors at Kenyon College, Gambier, Ohio, on May 23–24, 1969:

1. (a) Exhibit a countably infinite disjoint collection of infinite subsets of positive integers.

(b) Determine the cardinal number of the set of real-valued continuous functions on the interval [0,1].

(c) True or False: There exists an uncountable family F of distinct subsets of the positive integers such that for every $A, B \in F$ either $A \subset B$ or $B \subset A$.

2. Give an example of a closed differential form which is not exact.

3. (a) Define the characteristic of a field, and show it must be prime.

(b) Show that the order of a finite field must be a power of a prime.

(c) Construct a field of 25 elements.

(d) Is there an uncountable field of characteristic 2?

4. For $0 < t < 2\pi$, let f_t be the characteristic function of the interval $0 < x < t$. Regarding f_t as an element of the Hilbert Space L_2 on $[0,2\pi]$, consider its Fourier coefficients with respect to the complete orthonormal set $\{e_n\}$, where

$$e_n(x) = e^{inx}/(2\pi)^{1/2}, n = \ldots,-2,-1,0,1,2,\ldots .$$

(a) Use Parseval's equation to evaluate

$$\int_0^{2\pi} \cdots + |\int_0^t e_{-1}(x)\, dx|^2 + |\int_0^t e_0(x)\, dx|^2 + |\int_0^t e_1(x)\, dx|^2 + \cdots dt,$$

almost without computation.

(b) Use the result of (a) and a little computation to derive Euler's formula $1 + \frac{1}{4} + 1/9 + 1/16 + \cdots + 1/n^2 + \cdots = \pi^2/6$.

5. (a) State or derive the general form of a conformal mapping of the open unit disk onto itself.

(b) If $f(1/2) = 0$ and $|f(z)| \leq 1$ for $|z| \leq 1$, show that $|f(-1/2)| \leq 4/5$. (f is analytic in the disk.)

6. (a) Find the maximum value of xyz^2 in the positive octant, subject to the constraint $x + 2y + 3z = c$, a constant, by means of Lagrange multipliers.

(b) State the inequality of the geometric and arithmetic means, giving due attention to the case of equality.

(c) Solve (a) by use of (b).

7. Let X be a metric space with metric d; let A be a compact connected subset of X, and p a point of $X - A$. Suppose that $f:A \rightarrow A$ is continuous.

(a) Show that there is a point a in A such that $d(a,p) = d(f(a), p)$.

(b) What if the compactness of A is dropped?

(c) What if the connectedness of A is dropped?

8. Let $(Q,+)$ denote the additive group of the rational numbers.

(a) Show that $(Q,+)$ is not a free Abelian group.

(b) Show that every finitely generated subgroup of $(Q,+)$ is free Abelian.

9. (a) Show that every compact metric space is separable.

(b) What is the largest cardinality that a separable Hausdorff space can have? Hint: Consider the closures of open sets.

10. (a) Let V be the vector space of all real-valued continuous functions on the reals. Let $T:V \rightarrow V$ be defined by

$$(Tf)(x) = \int_0^x f(t) \, dt.$$

Prove that T has no characteristic values.

(b) Let W be the vector space of polynomials with real coefficients and with degree 3 or less. Define $S:W \rightarrow W$ by $S(p) = p'' + p' + p$, where primes denote differentiation as usual. Find the characteristic and minimal polynomials of S. Is S diagonizable?

11. Prove that no infinite-dimensional Banach space is locally compact.

12. (a) Use your knowledge of the homology groups of cells and spheres to show that there is no continuous function $f:B^n \to S^{n-1}$, where B^n is the unit ball in R^n and S^{n-1} is its boundary, having the property that $f(x) = x$ for every x in S^{n-1}.

(b) If $\pi_1(X,x_0) = A_5$, the alternating group on 5 elements, what can you say about $H_1(X)$ (integral coefficients)?

13. With the usual inner product on R^3 obtain an orthonormal basis from $v_1 = (1,2,2)$, $v_2 = (2,1,-2)$ and $v_3 = (2,4,-5)$ by means of the Gram-Schmidt process.

14. Evaluate by the residue calculus $\int_{-\infty}^{\infty} e^{itx}(1 + x^2)^{-1}\,dx$.

15. Use the fact that the real number system is an Archimedean ordered field to prove that between any two real numbers there lies a rational number.

16. If X is compact and connected, show that any two points in X lie in a minimal (i.e. irreducible) compact connected subset.

17. (a) By using Zorn's lemma, prove that in any vector space a linearly independent set of vectors can be extended to a basis.

(b) Discuss the functional equation $f(xy) = f(x)f(y)$, for f an unknown function from $(0, \infty)$ to $(0, \infty)$.

18. (a) Let V denote the vector space of polynomials in one variable over R, the real numbers. For x_0 a real number, define the linear functional $L(x_0):V \to R$ by $[L(x_0)](p) = p(x_0)$.

(a) Show that $\Lambda = \{L_x \mid x \in R\}$ is linearly independent.

(b) Find a linear functional on V which is not in the space spanned by Λ.

If this examination were given today, after fifty years of innovation, there would be no honors mathematics graduates in the United States.

inscribe The Latin verb *inscribo, inscribere, inscripsi, inscriptus* means *to write (scribo) on (in)*.

instantaneous The Latin suffix *-aneus* is added to nouns to produce adjectives with the meaning of *pertaining to*. The Latin verb *insto, instare, institi* means *to stand on, to be close to, to follow closely*. Its present participle *instans, instantis* has the meaning *present, urgent*; its stem is *instant*, which became an English word. There was never a Latin adjective *instantaneus*. Nevertheless, the word is formed on good principles. The English adjective *instantaneous* is modeled on the analogy of *simultaneous* coined in the seventeenth century on the analogy of *momentaneous*, from the Latin adjective *momentaneus,-a, -um, lasting but a moment*. There never was a Latin word *simultaneus*.

integer This is a Latin adjective meaning *whole*. Thus, *numeri integri* means *the whole numbers*.

integral The Latin adjective *integer* means *whole*. By the sixth century A.D. the Latin adjectival suffix *-alis* was superimposed on the stem to form the medieval Latin adjective *integralis, -e*, with the meaning *pertaining to the whole*. From this there came into French and English the noun *integral*.

integrand From the adjective *integer*, which means *whole*, there was formed the first-conjugation verb *integro, integrare, integravi, integratus, to make whole*, whose gerundive is *integrandus, -a, -um*, meaning *[that] which must be made whole*, whence comes the English noun *integrand*.

integration From the stem of the fourth principal part *integratus* (*made whole*) of the verb *integro, to make whole*, there was formed the noun *integratio, integrationis, a making whole*, whence came the English word *integration*.

intercept The Latin verb *capio, capere, cepi, captus* means *to seize*. The addition of the Latin preposition *inter* as a prefix produced the verb *intercipio, intercipere, intercepi, interceptus*, which means *to take by the way*. The word *intercept* is the stem of the fourth principal part.

interest This is the Latin verb meaning *it concerns, it is a matter of concern to*. The basic facts are as follows: If a customer deposits *x* dollars at a bank that offers annual interest rate i, with the interest compounded n times during the year at the end of n equal time periods, then the amount of money in the account at the end of y years is given by

$$x[1 + (i/n)]^{ny}.$$

If the interest is compounded continuously, one takes the limit as n approaches infinity to conclude that the amount in the account has grown to xe^{iy}. Furthermore, if a fellow takes out a mortgage of amount M for y years from a bank that charges annual interest rate i compounded continuously, then the monthly payment P is related to the other parameters by the equation

$$P = [Me^{yi}(1- e^{i/12})] / (1 - e^{yi}).$$

interior This is the Latin comparative adjective meaning *inner*. The superlative is *intĭmus, innermost*.

intermediate The Latin adjective *intermedius, -a, -um* means *in the middle*; it is the combination of the preposition *inter, between*, and *medius*, which means *middle*.

interpolate The Latin preposition *inter* means *between*, and the verb *polio, polire* means *to polish, file, make smooth*. From these two words were formed the adjective *interpolis* with the meaning *furbished up, vamped up* and the verb *interpolo, interpolare, interpolavi, interpolatus* with the meaning *to furbish, vamp up*, and thereby *to falsify by making something look better than it is or by sticking something in*. From the last principal part of the verb came the English word *interpolate*.

intersect This is the stem of the fourth principal part *intersectus* of the Latin verb *interseco*, formed from the preposition *inter, between*, and the first-conjugation verb *seco, secare, secui, sectus, to cut*.

interval The Latin preposition *inter* means *between*; the noun *vallum* means *wall*. Combining the two produces the noun *intervallum, an intervening space*, whence comes the English word *interval*. An interval of real numbers is a set of all real numbers between two given real numbers; each of the given boundary points may be included in the interval or not. For example, the set of real numbers greater than or equal to *a* but less than *b* is denoted [*a*,*b*).

intrinsic The Latin adverb *intrinsecus* means *on the inside, inwardly*. It is formed by the amalgamation of the two adverbs *intro, inwards, within,* and *secus, otherwise.*

intuitionism The Latin verb *intueor, intueri, intuitus* means *to look at with attention.* From the third principal part came the noun *intuitio. intuitionis,* which means *gazing* and gave us the noun *intuition.* In the nineteenth century, it became the fashion in English to make nouns out of other nouns by adding the suffix *-ism*, from the Greek nominal ending -ισμός. The idea was to give a name to a movement somehow aptly described by the noun to which *-ism* is appended. Communism, nihilism, socialism and all the other *isms* came about in this way. Mathematical intuitionism is the school of thought that holds that mathematics is constructed, not discovered, by the mind.

invariant From the Latin verb *vario, variare, to diversify, change,* is formed the present participle *varians, variantis,* from which comes the noun *variantia, difference, variation.* By adding the negating prefix *in-,* one produces the noun *invariantia, lack of difference, lack of variation,* whence is derived the English noun.

inverse This adjective is formed from the fourth principal part of the Latin verb *inverto, invertere, inverti, inversus,* which means *to turn over, to turn* (*verto*) *on* (*in*) *its back, transpose.*

inversion The Latin verb *inverto, invertere, inverti, inversum* means *to turn over, to transpose,* and from its fourth principal part comes the noun *inversio, inversionis, transposition,* whence the English noun is derived.

involute The Latin verb *volvo, volvere, volvi, volutus* means *to roll or wind.* The addition of the preposition *in* as a prefix produces the compound verb *involvo, involvere, involvi, involutus* meaning *to wind on or* *around, to wrap up.* If a string is wrapped around a curve C, or if a string already wrapped around C is unwound tautly, then the locus of any point P fixed on the string is called an *involute* of C. Alternatively, if a tangent line rolls without slipping along a curve C, the *involute* of that curve is the locus of a fixed point P on the rolling tangent line. The theory of involutes was first studied by Huygens (1673). According to Struik (p. 264) and the *Oxford English Dictionary*, the name *involute* was first used for these curves by Charles Hutton in his *Mathematical Dictionary*, London, 1796.

involute of the catenary The parametric equations of the *involute of the catenary* are $x = t - tanh\ t$, $y = sech\ t$. This curve is also called the *tractrix* or *tractory*. The beagle Chipper at *(0,1)* is connected to her master at the origin by a taut leash; we assume that both are point masses, so the leash is of length *1*. The master proceeds to walk along the positive x-axis, and since Chipper is recalcitrant, he cruelly drags her along. If we ignore friction, the path that Chipper traces out as she is dragged along the ground is the involute of the catenary. The involute of the catenary is also the *curve of pursuit*. Suppose that a rabbit at the origin sees Chipper at *(0,1)* and immediately races out along the positive x-axis with constant speed *1*. Chipper at the same time pursues the rabbit in such a way that she is always aiming at the rabbit and always at a distance *1* from him. Then the path of Chipper is the involute of the catenary.

involute of the circle The *circulus involutus* (wound circle) is a spiral obtained by wrapping a string around a circle (or, equivalently, from unwinding a string from around a circle), a piece of chalk having first been attached to the endpoint of the string. The locus of the endpoint is the *involute*. The curve was first studied by Huygens in 1693. If the circle has center at the origin and radius r, and if the fixed point (the chalk) is at *(r,0)* before the unwinding commences, then the parametric equations of the involute are $x = r(cos\ \theta + \theta\ sin\ \theta)$ and $y = r(sin\ \theta - \theta\ cos\ \theta)$, where the parameter θ is the angle that the circle has been unwound.

involution The Latin verb *involvo, involvere, involvi, involutus* means *to roll (volvo) in or on (in), to envelop, to wrap up*. From its fourth principal part comes the noun *involutio, involutionis, envelopment*, from whose stem comes the English noun.

irrational This is the Latin adjective *inrationalis, without reason*, with the ending *-is* dropped and the *n* assimilated to *r*. It is the Latin translation of the Greek ἄλογος, *unreasonable, not having a ratio*, for the Greek noun λόγος means *word, reason, ratio*.

irreducible The Latin verb *duco, ducere, duxi, ductus* means *to lead*, and the *dux* is *the man who leads, the leader*. The Latin adjective *inredux, inreducis* means *not leading back*. The addition of the suffix *-abilis* to the stem produces another adjective meaning *not capable of being led back*. From this is derived the English word *irreducible*.

-ism This suffix is derived from the Greek -ισμός, which forms nouns of action from verbs ending in -ίζειν.

isochrone The curve of equal (ἴσος) time (χρόνος) is the plane curve along which two point masses falling under gravity and without friction will reach the bottom in equal times no matter from which different higher-up points they start from rest. It is the same as the *tautochrone, q.v.*, and the only curve with this property is the cycloid.

isogonal The Greek adjective ἰσογώνιος means *equiangular*; it is the combination of the Greek adjective ἴσος, *equal*, and the noun γωνία, *corner, angle*. Someone added the Latin adjectival ending *-alis* to form the adjective *isogonal*, though one would have expected *isogonial*, unless there was some confusion with the related noun γόνυ, γόνατος, *knee*. This happens often. The resulting concoction would originally have been comical to the learned, but now, like many mistakes, it has been sanctioned by immemorial usage.

isolate From the Latin noun *insula, island*, there came the Italian noun *isola* with the same meaning. From the latter noun came the verb *isolare*, from whose past participle *isolato* came the English verb *to* *isolate.* This word was considered unworthy of inclusion in *Johnson's Dictionary* (1755).

isometric This is the combination of the Greek adjective ἴσος, *equal*, and the adjective μετρικός, *pertaining to a measure*, from the noun μέτρον, *measure*.

isometry This is the English form of the make-believe Greek noun ἰσομετρία, formed correctly on the analogy of Greek noun formation from the adjective ἴσος, *equal*, and the noun μέτρον, *measure*.

isomorphic This is the English form of a pretended Greek word ἰσομορφικός, constructed from the adjective ἴσος, *equal*, and the noun μορφή, *shape*, with the addition of the adjectival ending -ικός. See the entry **homomorphism**.

isomorphism Greek nouns ending in the suffix -ισμός came into Latin ending in *-ismus* and then into English ending in *-ism*. This word, however, is a modern invention, used by Burnside in 1897 in his *Theory of Groups*.

isoperimetric This adjective was formed by the combination of the Greek ἴσος, *equal*, and περιμετρικός, *pertaining to the perimeter* (περίμετρον). See the entry **Dido**.

isosceles This is the word ἰσοσκελής, ἰσοσκελές that Euclid used to describe triangles with two but not three equal (ἴσος, *equal*) sides (σκελή, *legs*). It was taken over by transliteration into Latin by Boëthius when he translated the beginning of Euclid's *Elements* around A.D. 500, and from thence it eventually entered English.

isotropic This adjective was composed in modern times by taking the two Greek words ἴσος, *equal*, and τροπικός, *pertaining to a turn or turning* (τροπή).

iterate The Latin adverb *iterum* means *again*. From the fourth principal part of the related verb *itero, iterare, iteravi, iteratus* meaning *to do a second time, to repeat*, comes the English verb *iterate*.

-ive The Latin suffix *-ivus* (or *-tivus*) was added to verb stems to produce verbal adjectives "expressing the action of the verb as a quality or tendency" (Allen and Greenough, §251, p. 152). Such is the origin of the endings *-ive* and *-tive* in English. Thus, *dissipative* means *tending to dissipate*.

-ize The Greek ending -ίζειν was added to a noun stem to produce a corresponding denominative verb. This ending became *-ize* when the word was taken over into the English language or when an English word was coined on this model.

J

j The Latin *J, j* came in late as a fancy *I, i* to be used for beauty's sake when the *i* was a consonant or, in Italian, when the *i* ended a word, particularly if it was preceded by another *i*.

Jacobian The Seventy transliterated the Hebrew proper name יעקב into Greek by Ἰάκωβος, which in turn was transliterated into Latin by *Iacobus* or *Jacobus*. It corresponds to the English *Jacob* or *James*. The adjective *Jacobian* refers to the mathematician Jacobi (1804–1851).

join The English verb is formed from the Latin verb *iungo, iungere, iunxi, iunctus, to join*.

joint This is the corruption of the Latin past participle *iunctus* or *junctus, joined*, from the verb *iungo, to join*.

K

κ This is the Greek letter *kappa*, small case, and the symbol for absolute curvature.

K The use of capital *K* as an abbreviation for *thousand* is tacky and to be condemned. Similarly low class is the slang word *grand*.

kampyle The Greek adjective καμπύλος means *bent, curved*; it is derived from the verb κάμπτω, *to bend*. It is the name of a curve defined by Eudoxus (*fl.* middle of the fourth century B.C.) which is the locus of points satisfying the equation $x^4 = a^2(x^2 + y^2)$.

kilogram This is a word of the French revolutionary metric system adopted on April 7, 1795. It should have been spelled *chiliogram*, since it is derived from the Greek χίλιοι, -αι, -α, *a thousand*, and γράμμα, *a letter*. (Compare, for example, *chiliarch*, the commander of a battalion of one thousand men.)

kilometer This is another word of the French revolutionary metric system; it is one ten-thousandth of the distance from the north pole to the equator through the meridian of Paris. It should have been spelled *chiliometer*, since it is derived from the Greek χίλιοι, -αι, -α, *a thousand*, and μέτρον, *a measure, rule, or standard*.

kinetic From the Greek verb κινέω, *to move*, came the adjective κινετικός, *pertaining to motion*, whence we get the English adjective by removing the nominative case ending -ος. The English of the eighteenth century wrote *ck* for final *k* in words derived from Greek. In the nineteenth century the final *k* was dropped.

kurtosis This is a corruption of the Greek noun κύρτωσις, *a bump, convexity, the condition of being humpbacked, a camel's hump*, from the adjective κυρτός, *curved, bent*. The spelling *kurtosis* is wrong since *y* and not *u* is the proper transliteration of the Greek letter *upsilon*. One writes *Kyrie eleison* for κύριε ἐλέησον, not *Kurie eleison*. Thus, the word should be spelled *kyrtosis*. In the theory of probability, the *kyrtosis factor* is the fourth moment of the standardized random variable; it is usually denoted α_4.

L

lacuna This is a Latin noun with the meaning *cavity, hollow, dip*. It is used by textual critics to describe a situation where some words have fallen out of a text. The accent is on the middle syllable.

lacunary From the Latin adjective *lacunarius* meaning *having empty spaces* came the English adjective *lacunary*, by the dropping of the nominative ending *-us* and the change of the then final *i* to the better-looking *y*.

lambda The letter λ of the Greek alphabet is traditionally used for a characteristic value of a matrix. The capital *lambda*, Λ, is used as an artsy *A* by people not encumbered by any knowledge of Greek.

lamina This is the Latin word for a *plate*. The Latin plural, *laminae*, is, alas, not used in English; instead, it is fine to say *laminas*. The accent is on the first syllable.

large This word is from the Latin *largus, -a, -um*, which means *abundant, copious*. There is an interesting passage relating to the *law of large numbers* in the book *Religion and Science* by Bertrand Russell (Oxford University Press, 1961, p. 158):

> It is said (though I have never seen any good experimental evidence) that if you toss a penny a great many times, it will come heads about as often as tails. It is further said, that this is not certain, but only extremely probable. You might toss a penny ten times running, and it might come heads each time. There

would be nothing surprising if this happened once in 1,024 repetitions of ten tosses, but when you come to larger numbers, the rarity of a continual run of heads grows much greater. If you tossed a penny 1,000,000,000,000,000,000,000,000,000,000 times, you would be lucky if you got one series of 100 heads running. Such at least is the theory, but life is too short to test it empirically.

Actually, the law of large numbers assures us that if you want the probability to be at least 99% that the difference between heads and tails will be at least one million, it is enough to toss the penny 6,400,000,000,000,000 times. The probability of the event mentioned in the penultimate sentence is actually about 40%. The law of large numbers appeared for the first time before the learned world in Chapter IV of the posthumous *Ars Conjectandi* of Jakob Bernoulli (1713): Suppose we toss a fair coin n times. If X is the number of heads obtained in those n tosses, then, given $\varepsilon > 0$ the probability that $\frac{1}{2} - \varepsilon < X/n < \frac{1}{2} + \varepsilon$ tends to 1 as n tends to infinity. Furthermore, if $M > 0$, then the probability that $|X - (n - X)| > M$ also tends to 1 as n approaches infinity.

lateral From the Latin noun *latus* meaning *side*, there was formed the adjective *lateralis* with the meaning *on the side*. *Lateral* is therefore the Latin adjective with the nominative case ending *-is* omitted.

Latin square The adjective *latinus* means *belonging to Latium*, the region around Rome. See also the entry **square**. A Latin square is an $n \times n$ square array, each of the n^2 entries of which is chosen from n different symbols in such a way that each symbol appears exactly once in each row and exactly once in each column.

latitude From the Latin *latus, -a, -um, wide*, came the noun *latitudo* by the addition of the nominal suffix *-tudo*. From this word proceeded the English noun *latitude*.

latus rectum This is a Latin phrase meaning *perpendicular side*. The *latus rectum* of an ellipse, a hyperbola, or a parabola is the chord through a focus perpendicular to the major axis of the ellipse, the transverse axis of the hyperbola, or the axis of the parabola, respectively. The Latin plural, which in this case must be used, is *latera recta*. English plurals like *latus rectums* or *latuses rectums* are comical.

lemma This is the Greek word λέμμα meaning *a peel, husk, skin, or scale that is peeled off*, from the verb λέπω, *to strip off, to peel*. The Greek plural is λέμματα, *lemmata*, but the English plural *lemmas* is in this case sufficiently sanctioned by custom so as to be acceptable.

lemniscate The *lemniscate* is the locus of points in the plane, the product of whose distances from two fixed points is one-quarter the square of the distance between those fixed points. It has the shape of a ribbon, and therefore Jakob Bernoulli, who first studied it (1694), named it in accordance with the Greek word λεμνισκός, *a ribbon*. If one allows the constant product to be an arbitrary positive real number rather than the particular value mentioned above, the resulting curves are the sections of a doughnut (*torus*) and are called the *ovals of Cassini*. If the two fixed points are *(−a,0)* and *(a,0)*, and if the constant product of the distances is a^2, then the polar equation of the lemniscate is $r^2 = 2a^2 \cos 2\theta$. Fagnano (1682–1766), a self-taught mathematician who studied this curve, ordered that it be inscribed on his tombstone along with the inscription *Deo veritatis Gloria*, which means, *Glory to the God of Truth*. The area of the region enclosed by the lemniscate is $2a^2$. An attempt to calculate the length will lead to an elliptic integral. The lemniscate is the pedal curve of the rectangular hyperbola with equation $x^2 - y^2 = 2a^2$ with respect to its center. If the given hyperbola is not rectangular, then its pedal curve with respect to its center is called the *hyperbolic lemniscate*; if the given hyperbola has equation $x^2/a^2 - y^2/b^2 = 1$, then the corresponding hyperbolic lemniscate has equation $r^2 = a^2 \cos^2\theta - b^2 \sin^2\theta$. The area of the region enclosed by the hyperbolic lemniscate is $(a^2 - b^2)\arctan(a/b) + ab$. The pedal curve of the ellipse with equation $x^2/a^2 + y^2/b^2 = 1$ with respect to its center is the *elliptical lemniscate*, whose equation is $r^2 = a^2 \cos^2\theta + b^2 \sin^2\theta$. The area of the region enclosed by the elliptical lemniscate is $\pi(a^2 + b^2)/2$.

lens The Latin word *lens, lentis* means a *lentil*. A lens is the plane region in common to two congruent circles whose intersection is not empty.

level The Latin adjective *lēvis* means *smooth*. It is to be distinguished from the adjective *lĕvis*, which means *light*. To its stem was added the diminutive suffix *-ellus* to form the adjective *levellus*; when the latter was shortened by the removal of the nominative singular ending *-us* so as to enter English in the usual manner, the extra *l* was also removed.

lexicographic From the Greek noun λέξις, λέξεος, *speech*, there was formed the adjective λεξικός, *pertaining to speech*. From the Greek noun γραφή, *drawing, writing*, came the adjective γραφικός, *pertaining to writing*. From the combination of the two was produced the English adjective *lexicographic*. The concoction *lexicographical* is wrong since it superimposes the Latin adjectival ending *-alis* upon the already present Greek adjectival ending -ικός. The word *lexicographic* is already an adjective and needs no further treatment.

liberal education The Greek word for *education* was παιδεία, the rearing of children (παῖδες), which came to be the Platonic technical term for the perfection of the mind, just as health (ὑγίεια = *salus*) is the perfection of the body, and virtue (ἀρετή = *virtus*) the perfection of the soul. The teacher was ὁ παιδαγωγός, the *child-leader* or *pedagogue*. The Greeks held that there were seven liberal arts, of which four formed the quadrivium of mathematics. The Romans termed *liberal* that education suitable for a free man, *homo liber*. The education of slaves, on the other hand, when undertaken at all, consisted of purely vocational matter. The classic treatment of the philosophy of a liberal education is by John Henry Newman (1801–1890); it was presented in a series of nineteen discourses written from 1852 to 1858. The collection was later published under the title *Idea of a University*.

limaçon This French word is derived from the Latin *limax, limacis*, which means *snail*. It is the locus of all points in the plane arrived at by the following process: Let C be the circle in the polar plane whose equation is $r = a \cos \theta$, and let the positive parameter b be given. On each line ℓ through the origin, mark off the points P and P' that are at a distance b on each side of C from the point of intersection of ℓ and C. Then the locus of P and P' is the *limaçon of Pascal*, defined by that authority in 1650 and given its name by Roberval. The polar equation is $r = a \cos\theta + b$. If $b > a$, the shape is that of a lima bean, or, in the opinion of those who are expert in such matters, of a shell-less snail, the slug. If $b = a$, the limaçon is a *cardioid*. If $b < a$, there is a baby inner loop. The limaçon is the conchoid of the circle C with respect to the pole. The area enclosed by the limaçon (first case) is $\pi(a^2 + 2b^2)/2$; the area enclosed by the small loop in the third case is

$$[(a^2 + 2b^2)(\pi - \arccos(-b/a))/2] \; - \; 3b(a^2 - b^2)^{1/2}/2.$$

The length of the limaçon is not expressible in closed form except in the case of the cardioid, when it is *8a*. The limaçon is the pedal curve of a circle with respect to any point. The cardioid is the epicycloid of one cusp. If $b = a/2$, the limaçon is called the *trisectrix* because by means of it one may trisect an arbitrary angle.

limit The Latin noun *limes, limitis* means a *cross path* or a *by-way*, *path* in general, and *boundary path* in particular.

line The Latin word *linea* means *line*. It translates the Greek γραμμή, and like that word can mean what we would call a *plane curve* or *arc*. If there was any ambiguity, one added the adjective ὀρθή, *recta*, straight.

lineal The Latin adjective *linealis* means *consisting of lines*. It is formed by the addition of the adjectival suffix *-alis* to the stem of the noun *linea*, line.

linear This is the Latin adjective *linearis*, *belonging to lines*, from *linea*, line. An equation of the form $y' + p(x)y = q(x)$ is called a *first-order linear differential equation*; it is linear because it is linear in y and y'.

liter The Greek λίτρα was *a unit of weight, a pound*, which came into late Latin as *libra, a balance or pair of scales*. At the time of the adoption of the metric system in France on April 7, 1795, *la litre* was defined to be a unit of volume, *viz.*, that of a kilogram of water at four degrees centigrade.

literal To the stem of the Latin noun *littera* or *litera, letter*, was added the adjectival suffix *-alis* to form the adjective *literalis*, which came over into English, upon removal of the case ending *-is*, as *literal*.

lituus The Latin word *lituus* was the *curved staff* of a Roman augur, or the *curved trumpet* of the Roman cavalry. Cotes in 1722 gave this name to the trumpet-shaped curve whose polar equation is $\theta = 1/r^2$. It was the symbol of the Roman College of Augurs, and may be seen on the coinage of those Roman emperors (for example, Nero) who had been admitted to membership in that sacred college; see Stack's catalogue of May 3–4, 1984, page 100, for illustrations. The crozier of bishops is just a development of the crooked augur's staff. The trumpet was named after the staff because of its shape.

local From the Latin noun *locus*, which means *place*, was formed the adjective *localis, belonging to a place*, by addition of the suffix *-alis* to the stem.

locus This is the Latin word for *place*. When one defines a curve as in the statement "The locus of points satisfying the equation…," it is equivalent to *set*.

log This is an abbreviation for *logarithm*. It should be used only in formulas, not in prose. It is also modern computer jargon; *to log on* means to enter one's "user name" and password into the computer so that it will work. In this second case the word is derived from the Arabic لوح, *a plank*, which acquired the meaning of a book in which one recorded one's progress in an enterprise; the related Hebrew word is לוּחַ, *a tablet*, for example, of the Ten Commandments.

logarithm The modern word λογαριθμός was formed by the happy union of the Greek nouns λόγος, *word, reason, ratio,* and ἀριθμός, *number.* It is an invention of Napier (1614).

logarithmic This is formed in accordance with the Greek rules for making adjectives, by adding the suffix -ικός to the stem of the modern invention λογαριθμός.

logic The Greek adjective λογικός, *pertaining to speech or to reason,* is derived from the Greek noun λόγος, *word, reason,* to whose stem the adjectival suffix -ικός has been appended. English then drops the last syllable -ός.

logistic The Greek noun λογιστής, *one who calculates or computes,* is derived from the deponent verb λογίζομαι, *to calculate or compute.* From the noun λογιστής, there was formed the adjective λογιστικός, *pertaining to calculation or the calculator,* by the addition of the usual suffix -ικός. These words are all related to λόγος, *word,* and to λέγω, *to speak.*

lognormal This modern concoction was formed from the juxtaposition of the stem of the Greek noun λόγος, *word, reason,* and the stem of the Latin adjective *normalis, pertaining to the carpenter's rule,* the *norma.* It was therefore originally a low, macaronic word. It is the name of the probability distribution of the random variable Y defined by $X = ln\ Y$, where X has the standard normal distribution.

-logy The Greeks appended the suffix -λογία from λόγος, *word, reason,* to certain words to indicate *the study of;* for example, θεολογία, from θεός, *god,* meant *the study of things divine.* Modern words formed on the analogy of this construction, such as *geology* and *topology,* are unobjectionable. Incorrect formations, however, are words like *sociology,* where the Greek suffix is appended to a Latin stem. Outright contemptible are things like *cosmetology* and *mixology,* which seek to throw the mantle of learning over low activities.

long This is the stem of the Latin adjective *longus.*

longitude The Latin word for *length* is *longitudo, longitudinis*.

loxodrome The Greek adjective λοξός means *slanting, crosswise, oblique*, while the noun δρόμος means *a course, a race*; the latter noun is related to δραμεῖν, the second aorist infinitive of τρέχω, *to run*. This is a modern word for a curve on the sphere that intersects each meridian of longitude at the same angle. There was already a classical Greek word for *oblique-running*, λοξοτρόχις.

loxodromic This word is formed by the addition of the stem of the Greek adjectival suffix -ικός to the stem of the modern noun λοξόδρομος. It means *pertaining to the loxodrome*. Loxodromic mappings are discussed by Konrad Knopp in *Elements of the Theory of Functions*, translated by Frederick Bagemihl, Dover Publications, Inc., 1952, page 58.

lune The Latin word for the moon is *luna*. A *lune* is the region of a circle remaining after the removal of a *lens*.

M

magic The Greek adjective μαγικός means *pertaining to a* μάγος *or Persian wise man*. The Greek adjective was transliterated into Latin as *magicus*, and from that, upon removal of the case ending, we get our noun *magic*.

magnitude The abstract Latin noun *magnitudo*, which means *great size*, is derived from the adjective *magnus, great*, by adding the nominal suffix -*tudo* to the stem.

major This is the Latin comparative adjective *maior*. The positive degree is *magnus*. The superlative is *maximus*. It means *bigger, greater*.

According to Struik (pp. 90, 265), the symbol > for *greater than* was first used by Harriot (1560–1621).

mantissa The Latin noun *mantissa* means *an addition of comparatively small importance, a makeweight*. In mathematics it is the decimal part of the logarithm, of less importance than the characteristic.

marginal The Latin noun *margo, marginis* means *border, edge*. The adjective *marginalis* is formed by addition of the adjectival suffix *-alis* to the stem of the noun.

mass The Latin noun *massa* means *a lump*.

math This is a regrettable abbreviation, no less silly than it is natural, for *mathematics*. Underwood Dudley was correct when, while lecturing on his book about mathematical quacks, he condemned the use of this word as opposed to the dignity of the subject. It is an example of *cataloguese*, like *chem, comp sci, poly sci, psych*, and so on. Some subjects escape this degradation, like philosophy, physics, and the names of languages. Cataloguese is a debased form of English and is extremely ugly. Its common use is deplorable.

mathematics This is the Greek word μαθηματικά, from the verb μανθάνω, *to learn*. It is the name of the subject whose branch of knowledge is traditionally and correctly equated with learning itself.
  The story is told by Vitruvius (*De Architectura* l. 6, c. 1) that once upon a time, the Socratic philosopher Aristippus was travelling by sea with some of his disciples when a storm arose, and they were shipwrecked on an island. The students were alarmed and expressed their concern that the inhabitants might lay violent hands upon them. Amid the panic, Aristippus took a look around and then commented that they had no reason to fear because he had just noticed the signs of intelligent humanity on the beach. The students humbly asked, "Where are those signs?" The philosopher pointed to a diagram that had been drawn in the sand with a stick. It was the diagram for Proposition 1 of Book I of Euclid's *Elements of Geometry*, the construction of the equilateral triangle. What was the lesson? It was

that the infallible and reassuring sign of civility is mathematics. Mathematical people are not dangerous, and they may be expected to behave rationally.

Gauß called mathematics *the Queen of the Sciences*. As we see from the story of Aristippus and from the title assigned to mathematics by Gauß, our subject has always been acknowledged to be not only special, a sign of civilization, but paramount, a sovereign. Statements to this effect by famous thinkers may be multiplied without end. The chief personality of the Enlightenment had this to say about mathematics in his *History of the Decline and Fall of the Roman Empire*:

> The mathematics are distinguished by a peculiar privilege, that, in the course of ages, they may always advance and can never recede. (Chapter LII, p. 427 of the first edition of vol. 5, 1788)

Thus, in the opinion of Gibbon, our discipline is associated with progress, a thing that humanity respects and desires. Because of this progress, it is responsible for our quality of life. In our scientific age, mathematics leads to success, and Americans worship success. It is a time-tested means to develop the mind and has always held a high place in *liberal education*, the system of general instruction that aims not to prepare the student for employment, but rather intends to free him from the handicap of intellectual slavery, that is, from a helpless dependency on others in the great questions of life. At one time, the branches of mathematics were considered to form four of the seven liberal arts, the *quadrivium*, while everything else was assigneded to the less important *trivium*, whose inferior status is reflected in the modern meaning of the word *trivial*.

Everyone knows that an athlete who strives to perfect his physique uses a specific exercise to bring out each muscle. The curl develops the biceps, the bench press works the chest, running strengthens the heart, and if he wants big thighs, he does squats. The athlete looks to see which body parts are strong and which are relatively weak, and then determines what exercises to do in order to build up the weak parts. Conversely, if one sees someone with very big triceps, one may rightly conclude that in his spare time he does dips. From the development of a specific muscle, one deduces that a specific exercise has been used in the training. If one observes that a fellow is scrawny and weak, one must suppose that he has never entered the weight room. If he is fat, you know that he eats too much.

The idea may now occur to a curious fellow that perhaps the mind is developed in the same way, that there are certain exercises to be done, that is, certain subjects to be studied, in order to bring out various aspects of it, and that, furthermore, by listening to an individual in conversation, or by reading his books, one may easily discover what branches of learning he has made the special object of study.

One may reasonably choose as the subject of meditation the topic, *What is the function of mathematics in life?* Every branch of learning has some function. For example, William James wrote that the function of religion was to make the inevitable tolerable. One may also consider some related questions such as: What effect does the study of mathematics have on the mind? What are the noticeable signs in our conduct that indicate to the world that we have studied mathematics? Conversely, how can we infallibly determine, upon coming into contact with an individual, that he is a mathematician?

One may mention here that at one time there were people who imagined that the extensive and successful study of mathematics actually resulted in a physical alteration in the shape of the skull, that just as exercise results in hypertrophy, the growth of the muscle, so mathematical activity produces the *mathematical bump* on the brain. This hypothesis was discussed and rejected by Jacques Hadamard in his book *The Psychology of Invention in the Mathematical Field*, so one need not worry further about it. It obviously arose from analogy to bodybuilding, where each exercise affects a special muscle. No, one cannot tell whether someone is a mathematician merely by looking at him, although one is tempted to think so when one studies the Hals portrait of Descartes in Copenhagen, the best painting of a mathematician in existence.

In order to prepare to answer the questions raised a moment ago, it is necessary to review what we know from experience about mathematics and mathematicians.

All mathematics begins with definitions, so the first observation to be made about a mathematician is that he is the sort of person who requires that all technical terms be defined, and that all common words be used properly in accordance with their accepted meanings. The importance, indeed, the necessity of making the correct definitions is the subject of some of the most famous Platonic dialogues. Mathematicians, when they go out into the world, are therefore by training bothered by the wrong use of words. This peculiarity we share with the learned of other professions. The wrong use of words causes us to pause and wonder, and is therefore a waste of our time. Now if we coin new words ignorantly or use old words incorrectly, we are not understood, or what we have to say is ugly to hear. Consider the vocabulary of technology: Mozilla Firefox, Skype, Nero Showtime, Safari, the Geek Squad, WeBWorK, LaTeX, and so forth; intelligence is no help in making sense out of such words. They are cant, which cannot be interpreted by the laws of etymology or by the usage of the best authors.

A mathematician requires that all assumptions be announced before the beginning of an argument and that they not be contradictory. Bertrand Russell has observed that the medieval theologians, who took their model from Euclid, were experts in deductive reasoning; when they went wrong, it was usually in their assumptions. The Founding Fathers showed the influence of mathematics when they wrote, "We hold these truths to be self-evident..." They were stating their axioms; they were in search of Euclidean axioms in politics. (See his *History of Western Philosophy*, The Folio Society, London, 2004, p. 36.) Where *they* went wrong was that they did not implement the theorems that followed from their postulates. They declared all men to be created equal, but they kept their slaves. There is nothing more unmathematical than to hold that something is correct in principle, but that it cannot be put into practice. If something is true, the mathematician naturally wants to implement it. This was an observation of the utilitarian economist James Mill, who had studied mathematics extensively in his youth.

A mathematician always requires proof of any claim presented for his acceptance. Therefore, a fellow who has studied mathematics successfully is expected to be capable of reasoning correctly and with clarity from appropriate assumptions. This is called *consecutive thought* or *deductive reasoning*, and not everyone is capable of it. The mathematician's demand that each statement be supported by a proof means that the study of mathematics, particularly if pursued exclusively, leads to skepticism. The most famous example of skepticism was Descartes. In philosophy he swept away the rubbish of his predecessors, though Will Durant thought that he then replaced it with his own rubbish. Mathematicians will be suspicious of what they are told if the claims made to them are not supported by a demonstration. If you tell a mathematician that the babies Romulus and Remus were fed by a woodpecker, or that St. Patrick lit a fire with an icicle, he will suppose that such events were unlikely to have happened.

Those people with a special gift for mathematics may be identified by their fascination with deductive reasoning. The following story is taken from Aubrey's *Life of Thomas Hobbes*:

> He was 40 years old before he looked on geometry, which happened accidentally. Being in a gentleman's Library, Euclid's Elements lay open, and 'twas the 47 El. Libri I. He read the proposition. By G-d, sayd he (he would now and then sweare an emphaticall Oath by way of emphasis), this is impossible! So he reads the demonstration of it, which referred him back to such a Proposition; which proposition he read. That referred him back to another, which he also read. *Et sic deinceps* that he was demonstratively convinced of that trueth. This made him in love with Geometry. (John Aubrey, *Brief Lives*, edited by Richard Barber, Woodbridge and Rochester, the Boydell Press, 1997, p. 152)

This is a story of which the Italians say, *Se non è vero, è ben trovato.*

The mathematician requires that the proofs that are presented for his inspection be written clearly. A poorly written proof will not work; it will confuse rather than convince the reader. Furthermore, an unclear demonstration alerts the reader to the intellectual limitation of its author, whom he will then distrust. The necessity to write and speak well means that it is important for a mathematician to be a master of the English language. This is difficult to accomplish today because the humanities are in free fall. Competence in the humanities is equivalent to writing well. Good writing is a result of years of reading the best authors. Today, though, everyone can read, but no one knows what is worth reading. People write poorly because they read rubbish. Computer science, technology, and certain sociological movements have had a catastrophic influence on the vocabulary of the English language. As a result, we live in a world where you must educate yourself in English, and to be self-educated is a disadvantage. What is more, many of the humanities long ago reached the limits of which the human being is capable and have been in decline. Is there any poet today who can instruct Homer or Dante, Vergil or Milton? Is there a general at the Pentagon who can write his memoirs like Julius Caesar? Is there a president who can write his own book like Marcus Aurelius? Where is the musician who surpasses Mozart, or the architect who leaves Michelangelo in the dust? Yet Johnny Average in any modern calculus course can tell Archimedes, Newton, or even Gauß a thing or two about mathematics that they would be astonished and grateful to hear. We live in a good time to study mathematics, and we live in a bad time to learn how to write English.

Observe also that a mathematician will react angrily, or at least indignantly, to the abuse of his subject by quacks. For example, social scientists attempt to compensate for the precariousness of their activities by throwing the mantle of mathematics, in particular of statistics, over their productions. We are constantly pestered with requests by telephone or email to fill out questionnaires in which we are to choose 1, 2, 3, 4, or 5 for our answer. The social scientists then analyze the data thereby collected and draw conclusions from the averages. It is a flight from reality into pseudoscience. Technically speaking, the problem is that it is impermissible to work with means of ordinal data.

Mathematicians are impatient with cranks. Such specimens waste our time. Cranks are everywhere; there is one behind every bush. The classic examples are the circle-squarers, cube-doublers, trisectors, and heptagon-constructors. It is the mark of a crank to set aside centuries of serious thinking on a subject. The ones mentioned above do so when they reject the theorems that prove that the constructions they attempt are impossible with unmarked straightedge and compass alone; the reason they do this is that they do not have the education required to understand the proofs. They are intrigued by the problems, which can be stated simply and whose formulations require no special learning to understand. But the quacks are lazy and assume that the alleged proofs are wrong, or mere opinion, or part of a conspiracy, and produce their own pretended solutions. These solutions are constructions in the style of a recipe, do this, do that, and they give no demonstration that the construction accomplishes what it claims to do. The world is full of such people. They have little understanding of consecutive argument, and dismiss the discoveries of science. Conversation with them is unproductive because they have no respect for learning. It is a mentality opposite to that of the mathematician. What the two types have in common is a passion, but in the circle-squarer this love is not followed by the appropriate education.

In order to understand the characteristic behavior of mathematicians, it is necessary to have some experience observing such people, preferably those of the top rank.

One may ask, Does mathematics make us better people in the ethical sense? I wish it were true, but I don't think so. Who dares to say that mathematicians are better people than historians or plumbers? There are many mathematicians of quality working night and day, even as I write, to give the atomic bomb to Kim Jong-un. Having a Ph.D. does not imply that one is nice. Joseph Goebbels had a doctorate in German literature from the University of Cologne. I recall that in the seventies, Nathan Jacobson told me that the mathematician Pontryagin was behaving disgracefully in an anti-semitic episode that was occurring in the Soviet Union at the time. Mathematics does not prevent you from doing stupid things. Galois got himself shot in a duel, a dumb thing to do. Mathematics doesn't require you to do "service learning" or community involvement. Mathematicians are not social workers or priests. I also do not make the claim, that mathematics makes us more active citizens. I doubt that this is true.

Mathematicians who are teachers regularly adopt the bureaucratic mentality as they function within their colleges and universities. By doing so they torpedo their own ship and demonstrate that they do not understand the meaning of their own profession. They thereby contribute to the decline in the estimation in which society holds their profession.

> In the last thirty years the prestige of mathematics has declined in all countries. I think that mathematicians are partially to be blamed as well (foremost Hilbert and Bourbaki), particularly the ones who proclaimed that the goal of their science was investigation of all corollaries of arbitrary systems of axioms. (Vladimir Arnold quoted in *Tribute to Vladimir Arnold* in the March 2012 issue of the *Notices of the American Mathematical Society*, vol. 59, no. 3, pp. 378–399)

What then is the function of mathematics in life? I say that the function of mathematics is to apply the method of deductive reasoning to the world around us and to the world in our minds. It is the key to understanding the Universe.

> Mathematics is the great conclusive example of the discovery of truths by reasoning. (John Stuart Mill, inaugural address as rector of the University of St. Andrews, 1867)

The objects to which we apply the method vary as the problems that we choose to examine. How can one tell that a fellow is a mathematician? The mathematician looks for and distinguishes between definitions, assumptions, and propositions; he uses words without ambiguity, recognizes axioms whether stated or hidden, and distinguishes between demonstrated propositions and claims for which no proof is provided. A mathematician does not accept appeals to authority as equivalent to a demonstration. What effect does the study of mathematics have on the individual? What are the habits he acquires? What corresponds in his personality to the "mathematical bump"? The mathematician is impatient with quackery, obscurantism, and pretence, and is skeptical, or at least careful, in matters of philosophy or belief. He holds fast to the consequences of any axioms he assumes. As a result of these professional habits, he is poor at compromise. He will not implement error upon command of authority. In life, he is not easy to use.

Why should one study mathematics? Aristotle defined man as the unique animal capable of reason and art. The study of mathematics inclines the student toward living his life in accordance with reason and art. If the muscles in the brain that control reason and art need to be developed, then the appropriate stimulation is mathematics.

The last paragraph of Hume's *Enquiry concerning Human Understanding* is famous:

> When we run over our libraries, persuaded of these principles, what havoc must we make? If we take in our hand any volume; of divinity or school metaphysics, for instance; let us ask, *Does it contain any abstract reasonings concerning quantity of number?* No. *Does it contain any experimental reasonings concerning matters of fact or experience?* No. *Commit it then to the flames: For it can contain nothing, but sophistry or illusion.*

By *abstract reasonings concerning quantity of number* he means mathematics. Mathematics has thus far succeeded in escaping the ruin that has befallen its relatives in the humanities and social sciences.

mathlete This is a comical word concocted to describe a fellow who wins a mathematical competition. Compare *prequel*, formed in the same manner by the illiterate.

mathophile This is another absurd word since there is no Greek word *math*. It is supposed to mean a *lover of mathematics*.

matrix This is the Latin word for *womb*. It is derived from the word *mater, mother*.

matroid This is a low word, the combination of the Latin stem *matr* from *mater, mother*, and the Greek suffix *-oid* from εἶδος, *shape*. If the particle *matr* were meant to be from the Greek word for *mother*, it ought to have been *metr-*, as in *metropolis*.

MatSciNet Scott Guthery is the author of the article "Google Books vs. MathSciNet" in the September 2012 issue of the *Notices of the AMS*, pages 1115–1116. He writes:

> MathSciNet is used here to refer specifically to the restricted-access database of telegraphic summaries of scholarly mathematics maintained by the American Mathematical Society.

In the same issue of the *Notices* there is mention of the *Mandelbulb* (p. 1061), the Mandelbrot set in 3D. In truth, neither Euclid nor Euler could have conceived of creating names like *MathSciNet* or *Mandelbulb*. This is only possible in an age where the English language is subconsciously viewed as a plaything, even in the learned world.

maximal The Latin adjective *maximus* is the superlative degree of *magnus*, *big*. Though already an adjective, its neuter singular form *maximum* acquired the force of a noun, *the maximum*, and so was able to suffer in English the addition to its stem of the adjectival suffix *-al* to produce the word *maximal*. There is no Latin word *maximalis*.

maximum This is the superlative degree, neuter gender, of the Latin adjective *magnum*, *big*.

mean The Latin adjective *medius* means *in the middle*. It is related to the Greek adjective μέδιος of the same meaning. In the Middle Ages, the adjectival suffix *-anus* was superimposed on the stem to produce another adjective *medianus* with the same meaning. The *d* has dropped out probably to conform to another word *mean*, related to the Latin *communis* and the German *gemein*. There is also a third *mean*, related to the German verb *meinen*. See Weekley, *sub voce*. The mean ergodic theorem of von Neumann (1932) says that if T is a measure preserving transformation on *[0,1]* and if U is a mapping from L_2 onto itself defined by $U(f(x)) = f(T(x))$ for all f in L_2, then for every f in L_2 there is an f^* in L_2 such that $[f + U(f) + U^{2}(f) + \cdots + U^{n-1}(f)]/n$ converges to f^* in the norm of L_2 as n approaches infinity.

measurable This is the metamorphosis of the late Latin adjective *mensurabilis*. The English word was produced by the addition to the noun *measure* of the English suffix *-able*, derived from the Latin suffix *-abilis*.

measure The Latin verb *metior, metiri, mensus* means *to measure*. From its third principal part was derived the noun *mensura* with the meaning *measuring*. The Latin verb is related to the Greek noun μέτρον, *measure, standard*.

mechanics The Greek noun μηχανή means *an instrument, a device, a contrivance, an artificial means of doing something, a machine*. Among the Greeks, neither the unmarked straightedge nor the compass was considered a μηχανή, but all other instruments for making curves were condemned as *machines*. The addition of the adjectival suffix -ικός to the stem produced the adjective μεχανικός, *pertaining to a machine*. Adding an *s* to the stem produces the English word in question. This is the same common process that gave us the words *dynamics, mathematics, metrics,* and *physics*.

median The Latin adjective *medius* means *in the middle*. It is related to the Greek adjective μέσος of the same meaning. The adjectival suffix *-anus* was superimposed on the stem of what was already an adjective to produce another adjective *medianus* with the same meaning.

medium The Latin adjective *medius, -a, -um* means *in the middle*.

member The Latin noun *membrum* means *a limb*.

mensuration This is the stem of the Latin noun *mensuratio, mensurationis*, which was formed from the third principal part of the verb *metior, metiri, mensus*, which means *to measure*.

mentee, mentor These are overused words nowadays; the former is also comical. In the *Odyssey*, Μέντωρ was the advisor of Telemachus, son of Ulysses. The ending *-ee* is the corruption of the French ending *é.* It should therefore not be added to a non-French stem unless the intention is to be ludicrous.

meridian The Latin noun *meridies, meridiei* is the union of *medius, middle,* and *dies, day.* The addition of the suffix *-anus* to the stem produced the adjective *meridianus, pertaining to midday.*

meromorphic This adjective is compounded of the Greek noun μέρος, *a part,* μορφή, *a shape,* and the adjectival suffix -ικός. See the entry **morphic**.

> A single-valued function shall—without regard to its behavior at infinity—be called meromorphic, if it has no singularities other than (at most) poles in the entire plane. (Knopp, Part II, p. 35)

See the entries **pole** and **singularity**.

metacompact The Greek preposition μετά means *beyond.* The word *compact,* however, is from a Latin root; see the entry **compact** above. This is therefore a low word, an example of the Greek prefix *meta-* being used indiscriminately to indicate a sort of supercompactness. The proper word would have been *transcompact.*

metadata I learned of this new low word when I read the article "Reading Beneath the Lines" by William Triplett on page D5 of the *Wall Street Journal* of October 12, 2011. Will Noel, curator of manuscripts and rare books at the Walters Art Museum in Baltimore, is there quoted as saying:

> The Archimedes palimpsest breaks down boundaries between disciplines. It contains history, philosophy and mathematics, and then all the latest technologies that were applied—the digital imaging, the metadata management—along with all the scholarship.

The word is just macaronic cant, the combination of the Greek preposition μετά and the Latin noun *data.* The Greek preposition has been loosely used to indicate something antecedent to and transcending whatever follows.

metaharmonic This word appears in the title of a recent book by Willi Freeden, *Metaharmonic Lattice Point Theory* (CRC Press, 2011). It is formed correctly from the obvious Greek parts.

metamathematics This twentieth-century noun is compounded of the Greek preposition μετά meaning *beyond* and the name of our subject, μαθηματικά, *mathematics*. According to Stoll, *Set Theory and Logic*, Dover Publications, Inc., New York, 1963, pages 403, 404:

> In brief, metamathematics is the study of formal theories by methods which should be convincing to everyone qualified to engage in such activities....To prove theorems about such theories—in particular, to attack the problem of consistency—Hilbert devised metamathematics.

meter The meter is the unit of measurement in the scientific system, defined by the French revolutionary government as one ten-millionth of the distance from the north pole to the equator through the meridian of Paris. The metric system was officially adopted by decree of the National Convention on the 18th Germinal, year III, that is to say, April 7, 1795. The word *meter* was taken from the Greek μέτρον, *measure, standard*.

method The Greek noun μέθοδος means *a following after*; it is compounded of the words μετά, *after*, and ὁδός, *road*.

metric To the stem of the Greek noun μέτρον, *measure, standard*, there was added the adjectival suffix -ικός to produce the adjective μετρικός, *pertaining to measurement*. Our word *metric* is the stem of this adjective.

metrizable Many Greek verbs derived from nouns end in -ίζειν. This ending became *-izare* when those words were taken over into late Latin. This is the origin of our English *-ize*. The ending *-able* is the transfiguration of the Latin adjectival suffix *-abilis*. However, there is neither a Greek verb μετρίζω nor a Latin verb *metrizo*; the English verb *metrize* is old but rare and means *to impose a meter on*. The macaronic word *metrizable* should therefore mean *capable of having a meter imposed [on it]*. The correct word for the idea of *capable of having a metric imposed on it* would have been *metricizable*.

mid- This prefix is an English substitute for the adjective *middle*. See the following entry.

middle This adjective is derived from the Latin *medius*, which means *in the center*. The *-le* comes from superimposing the remnant of the Germanic diminutive *-lein* on the Latin *mid*.

midpoint See the entries **mid-** and **point**.

mile This is the misspelling of the Latin noun *mille*, *a thousand*. The original mile was one thousand paces.

millennium This is a good Latin word meaning *a period of a thousand years* from *mille*, *a thousand*, and *annus*, *year*. Because English is not careful to pronounce a double consonant longer than a single one, the word is often misspelled *millenium*, although *anus* is the fundament. The year 2000 was widely presumed to be the first year of the third millennium A.D., as mathematicians were not sufficiently influential to succeed in pointing out that it was the last year of the second millennium.

million The progression in America is million (10^6), billion (10^9), trillion (10^{12}), quadrillion (10^{15}), quintillion (10^{18}), sextillion (10^{21}), septillion (10^{24}), octillion (10^{27}), nonillion (10^{30}), decillion (10^{33}), undecillion (10^{36}), duodecillion (10^{39}), etc.; that is, the words are formed from the Latin numerals, and if there is a need for larger numbers, they must be formed on this analogy by a Latinist. No decree of any authority can legitimize hybrid creations like *teratillion*.

minimal The Latin adjective *minimus* is the superlative degree of *parvus*, *small*. The neuter singular form acquired the force of a noun, the *minimum*, and then, though already an adjective, suffered the addition to its stem of the adjectival suffix *-al* to produce the word *minimal*.

minimax The prefix *mini-* is added to nouns to produce comical diminutive words. Such words are even more comical when they are not realized to be so by their authors.

minimum The Latin adjective *minimus, -a, -um* is the superlative degree of *parvus*, *small*.

minor This is the masculine and feminine gender of the Latin adjective for *less*. According to Struik (pp. 90, 265), the symbol < for *less than* was first used by Harriot (1560–1621).

minus This is nominative and accusative singular neuter gender of the Latin adjective for *less*.

minute The Latin verb *minuo, minuere, minui, minutus* means *to make smaller*. The noun and adjective *minute* proceeds from the fourth principal part.

mirror The Latin verb *miror, mirari, miratus* means *to gaze at*. The extra *r* was added a millennium ago in an age careless of detail.

mixed The Latin verb *misceo, miscere, miscui, mixtus* means *to mix*. Our verb *mix* comes from the fourth principal part.

mnemonic The Greek adjective μνήμων means *mindful*. The related adjective μνημονικός means *pertaining to memory*. There are mnemonic devices to help in memorizing the digits of γ, e, and π. Boyer's device for π is: How I want a drink, alcoholic of course, after the heavy lectures involving quantum mechanics.

mode The Latin noun *modus* means *measure* and then *manner*. The Latin noun came into French as *mode*, and *mode* came into England in 1066.

model The diminutive of the Latin noun *modus* is *modulus*. This became *modello* in Italian and *modèle* in French, whence it entered English.

modular The diminutive of the Latin noun *modus* is *modulus*. By the addition of the adjectival suffix *-aris* to the noun's stem, there was produced the adjective *modularis, having to do with measure*, which is the origin of the English word.

module The diminutive of the Latin noun *modus, standard of measurement*, is *modulus*. Modulus became *module* in French, and this is where we got the English word for *a small measure*. A *module* is a generalization of a vector space where the scalars are chosen from a ring instead of a field; the name is inappropriate for the object it denominates.

modulo This word is the ablative singular of the noun *modulus*. It was taken over as is from the Latin compositions of the mathematicians of an age gone by.

modulus The diminutive of the Latin noun *modus* is *modulus*. The word therefore means *a little measure*. The *modulus* of a complex number $a + bi$ is its magnitude $(a^2 + b^2)^{1/2}$.

modus ponens This is the axiom of logic that says that if s and t are statements, if s is true, and if $s \Rightarrow t$, then t is true. Those who cannot function with Latin words call it the *law of detachment*. The medieval Latin philosophic phrase *modus ponendo ponens* (of which *modus ponens* is an abbreviation) means *the method of affirming by means of affirming*.

modus tollens This is the axiom of logic that says that if s and t are statements, if $\sim t$ is true, and if $s \Rightarrow t$, then $\sim s$ is true. The medieval Latin philosophic phrase *modus tollendo tollens* (of which *modus tollens* is an abbreviation) means *the method of denying by means of denying*.

moment This is the stem of the Latin noun *momentum*. The Latin verb *moveo, movere, movi, motus* means *to stir, to set in motion*. From this verb was formed the noun *movimentum*, which was simplified to *momentum*, with the meaning *motion*. The English *moment* is just the Latin *momentum* with the case ending removed. Allen and Greenough write (§239, note):

> Of these endings, *-men* is primary; *-mentum* is composed of *men-* and *to-*, and appears for the most part later in the language than *-men*: as, *momen*, movement (Lucretius); *momentum* (later).

momentum This is a Latin word for *movement*. See the entry **moment** above.

monadic See the entry **polyadic**.

monic This is an absurd word. It appears to be the stem of a Greek adjective μονικός related to μόνος, *sole, alone*, but there is no such word μονικός. It was coined to indicate a polynomial with integral coefficients and leading coefficient *1*. It makes just as much sense to define a polynomial with integral coefficients with leading coefficient *2* to be *deuteric*, with leading coefficient *3* to be *tritic*, with leading coefficient *4* to be *tettartic*, with leading coefficient *5* to be *pemptic*, with leading coefficient *6* to be *hectic*, etc.

monodromy The Greek adjective μόνος means *alone*, and the noun δρόμος means *a course, a race, running*, from δραμεῖν, *to run*. The *monodromy theorem* is discussed by Konrad Knopp (I, p. 105):

> Let G be a simply connected region and $f_0(z) = \sum a_n(z - z_0)^n$ a regular functional element at the point z_0 of G. Then if $f_0(z)$ can be continued from z_0 along every path within G, the continuation gives rise to a function which is single-valued and regular in the entire region G.

Thus, the modern word *monodromic* is equivalent to *single-valued*.

monogenic This is a modern adjective; there is no Greek word μονογενικός. The Greek adjective μόνος means *alone*, and the noun γένος means *descent*. The ending *-ic* is probably meant to be the stem of the Greek adjectival suffix -ικός. The compound adjective μονογενής, *only begotten*, appears in the last sentence of the *Timaeus*. *Monogenic analytic functions* are a type of analytic function discussed by Knopp (II, p. 138) that are *generated by a single element*.

monoid This is the transformation of the Greek adjective μονοειδής, μονοειδές, which means *like one, simple, of one kind*. As a mathematical word, it is extremely bizarre. The Greek adjective μόνος means *alone*, and εἶδος means *shape*. *Monoid* is supposed to mean *semigroup*, but if one means *semigroup* one should just say *semigroup*. Schwartzman indulges in Olympian acrobatics to explain the appropriateness of the word for this meaning, but his attempt is futile:

> A monoid is the same as a semigroup. The name may be explained by noting that a monoid fulfills only one part of the definition of a group: a monoid is a set of elements and a binary operation that is closed and associative; however, there need be no identity element or inverse elements. (Steven Schwartzman, *The Words of Mathematics: An Etymological Dictionary of Mathematical Terms Used in English*, The Mathematical Association of America, 1994, p. 139)

I have seen other definitions of a monoid that require the identity element, but in this case as well the name remains unexplained.

monomial This is the case of a Latin adjectival suffix *-alis* being added to the corruption of a Greek word. The Greek adjective μόνος means *alone, sole*, and the noun νόμος means *portion, custom, law, mathematical term*. The correct combination of the two would be μονόνομος, *having one term*. The word should have gone into English as *mononome*, and the corresponding adjective should have been *mononomic*. Alas, one of the two *-no-* syllables was dropped, an *-i-* was inserted, and then a Latin suffix was appended, a true comedy of errors. Or, more likely, it was constructed on the analogy of the low word *binomial*, the *bi-* having been replaced by *mono-*.

monomorphism This word is a modern invention compounded of the prefix *mono-* from μόνος, *alone, sole*, and the noun *morphism, q.v.* A monomorphism is a group homomorphism that is one-to-one. Herstein uses the word *isomorphism* for such a function.

monotone The Greek adjective μόνος means *alone*, and the noun τόνος means *that which tightens or is tightened, a musical note*. The adjective μονότονος thus means *having one note*.

monotonic The addition of the adjectival suffix -ικός to the adjective μονότονος, *having one note*, produces the adjective μονοτονικός, *pertaining to what has but one note*, and the stem of this Greek adjective is the English word.

Monte Carlo This means *Mount Charles* in Italian. In mathematics, it is the approximation of a mathematical constant by experiment. Buffon was the first to do so, when he approximated π by the tossing of a needle. A plane floor is ruled with horizontal lines d units apart. The mathematician takes a needle of length ℓ, $\ell \le d$, and tosses it at random on the floor a total of n times. His work done, he observes that the needle intersected a line i times out of n. Since the probability of an intersection is $2\ell / \pi d$, we may approximate π by $2\ell n/id$. In 1850, Wolf claimed to have tossed a needle 5,000 times and to have obtained a value of 3.1516 for π. It is possible to simulate the needle-tossing experiment with computer-generated random numbers. In 1989, I compelled a student Shane Michael Fisher to simulate the Buffon experiment by computer; he "tossed" the needle 34,000,000 times and got the value 3.14150106 for π.

morphic *Morphic* is a modern invention derived by attaching the Greek adjectival suffix to the stem of the noun μορφή, *a form, shape, or figure*.

morphism The Greek noun μορφή means *shape, figure, appearance.* The suffix -ισμός, which went into Latin as *-ismus* and into English as *-ism,* was added to nouns or adjectives to indicate a party, doctrine, or theory suggested by the substantive to which it was appended. *Morphism* should thus mean *the theory of shape.* This was all Greek to those who just needed a word to describe a certain type of structure-preserving function.

motion This word is the stem of the Latin noun *motio, motionis,* which means *movement.* It is derived from the fourth principal part of the verb *moveo, movere, movi, motus,* which means *to move.*

multi- This prefix is derived from the Latin adjective *multus, -a, -um,* which means *many* or *much.* It can rightly be used only with words of Latin origin. The corresponding Greek prefix is *poly-.*

multifoil This word is the union of the Latin prefix *multi-, many,* and the noun *folium, leaf.*

> In mathematics, a multifoil is a plane figure made from adjacent congruent arcs of a circle placed around the vertices of a regular polygon; the figure is made up of many "leaves." Depending on the number of leaves it possesses, a multifoil may be known as a *trefoil, quatrefoil, cinquefoil, hexafoil,* etc. (Schwatrzman, *sub voce*)

Notice the inconsistency of language in the choice of the numerical prefixes.

multinomial This low word is concocted on the analogy of Newton's word *binomial* by putting together the Latin prefix *multi-, many,* the stem of the Greek noun νόμος, *law,* and the Latin adjectival suffix *-alis.* The correct word for this concept would have been *polynomic.*

multiple This is derived from the Latin adjective *multiplex, multiplicis,* which means *folded many times.* In late Latin, *multiplex* became *multiplus,* and this accounts for the absence of the *c* in *multiple.*

multiplicand This is the stem of the gerundive *multiplicandus, that which must be multiplied*, of the Latin verb *multiplico*. See the entries **multiply** and **multiple**.

multiplicity This is the Latin noun *multiplicitas, multiplicitatis*, formed from the fourth principal part of the verb *multiplico*. See the entries **multiply** and **multiple**.

multiplier The French verb *multiplier* rendered the Latin *multiplico*. *Le multiplieur* was *the fellow who multiplied*. This became the English *multiplieur* and the American *multiplier*.

multiply The Latin verb *multiplico, multiplicare, multiplicavi, multiplicatus* means *to fold (plico) or increase many (multi-) times*. It is derived from the adjective *multiplex*; see the entry *multiple*.

multiset This is a bad word because it does not suggest what it means. It is a set in which some elements occur more than once, for example, {*a,a,a,b*}. See the entries **multi-** and **set**.

multivariable This is put together from the Latin prefix *multi-, many*, and the adjective *variabilis, capable of being changed*, from *varius, manifold*.

mutual The Latin verb *muto, mutare* means *to move, shift,* or *change*. From this verb comes the adjective *mutuus*, which means *interchanged, reciprocal*, and the noun *mutuum*, which means *reciprocity*. The addition of the suffix *-alis* to the stem *mutu-* produces the adjective of relation *mutualis* with the meaning *reciprocal*, whence the English adjective is derived.

Mysterium Cosmographicum This is the title of a book by Kepler (1596) in which he presents a heliocentric model for the solar system based on the Platonic solids of the *Timaeus*. There are six concentric spheres, with the sun at the common center, one sphere for each of the six planets known at the time. Inside the largest sphere, on which the planet Saturn moves, there is inscribed a cube. The sphere of Jupiter is inscribed inside the cube. A tetrahedron is inscribed inside the sphere of Jupiter. Inside the tetrahedron is inscribed the sphere of Mars. Inside the sphere of Mars is inscribed a dodecahedron, inside of which there is described the sphere of the earth. Inside the sphere of the earth is inscribed an icosahedron, and inside the icosahedron is inscribed the sphere of Venus. Inside the sphere of Venus is inscribed an octahedron, and inside the octahedron is inscribed the sphere of Mercury.

N

natural number The Latin verb *nascor, nasci, natus* means *to be born*, and from the third principal part comes the feminine singular participle *natura* meaning *that which is about to be born* and, as a noun, *nature*. From *natura* is formed, by the addition of the suffix *-alis* to the stem, the adjective *naturalis* with the meaning *pertaining to nature*. These are the positive integers, the numbers created by God, according to Kronecker.

necessary The indeclinable Latin adjective *necesse* means *unavoidable*. Allen and Greenough write (*New Latin Grammar*, §250a):

> Adjectives meaning *belonging to* are formed from nouns by means of the suffixes *-ārius, -tōrius*

Thus we get the adjective *necessarius*, whence was produced *necessary*.

negation The Latin word for *denial* is *negatio, negationis*, from whose stem we get the noun *negation*. *Negatio* itself is from *nego, negare, negavi, negatus, to deny*.

negative The Latin verb *nego, negare, negavi, negatus* means *to say no*. Allen and Greenough write (*New Latin Grammar*, §251):

Adjectives expressing the action of the verb as a *quality* or *tendency* are formed from real or apparent verb-stems with the suffixes *-āx, -idus, -ulus, -vus (-uus, -īvus, -tīvus)* .

Thus we get *negativus* from *negatus*, and finally *negative* from *negativus*.

nephroid Aristotle uses the adjective νεφροειδής, *shaped like a kidney*. It is the juxtaposition of the Greek word for *kidney*, which is νεφρός, and εἶδος, which means *shape*. It is the name of the curve that is *kidney-shaped*, the epicycloid of two cusps. If the radius of the fixed circle is *a*, and the radius of the rolling circle is *b* with $a = 2b$, then the parametric equations of the resulting nephroid are

$$x = b(3 \cos t - \cos 3t)$$
$$y = b(3 \sin t - \sin 3t).$$

The length of the nephroid is *24b*. The area of the plane region it encloses is $12b^2\pi$.

nerve This is the Latin noun *nervus*, which means a *sinew* or *nerve*. It is related to the Greek noun νεῦρον of the same meaning.

-ness This English suffix is added to adjectives in order to make nouns. Thus *good + ness = goodness*. It is wrong to add it to adverbs. Thus, *wellness* is a low word. It is also wrong to add it to adjectives that themselves are formed from an already existing noun. Thus *healthiness* is a low word since there is already the adequate noun *health*. The macaronic addition of this suffix to words of Greek and Latin origin is not recommended.

nilpotent *Nil* is a short form of *nihil*, Latin for *nothing*. *Potent* is from *potens, potentis*, the present participle of *possum, to be able*. The Romans did not add *nil* as a prefix to adjectives to form new adjectives; *nilpotent* is thus not a Latin word, but an unlearned modern invention.

nodal *Nodus* is the Latin word for *knot*. The addition of the adjectival suffix *-alis* to the stem produces the adjective *nodalis*, whence we get the English noun upon removal of the case ending *-is*.

node This noun comes from *nodus*, the Latin word for *knot*.

Noli tangere circulos meos. These are the last words of Archimedes—*Get off my circles!*—addressed to a Roman soldier who was trampling on a mathematical diagram that he had just drawn with a stick in the sand. The scene is the subject of a famous mosaic preserved in the Städelsches Kunstinstitut und Städtische Galerie in Frankfurt am Main.

non- The Latin adverb *non* means *not*. The English habit of forming one word by adding *non-* to adjectives (for example, *non-singular*) instead of using two words (for example, *not singular*) is a habit not in conformity with the best style. The use of the connecting hyphen is recommendable to prevent an unseemly concatenation.

nonagon This is an absurd word used by the unlearned for *enneagon*, *q.v.* It is the same sort of concoction as *septagon*, *q.v.*

non-homogeneity See the entries **non-** and **homogeneity**. This is the property of a polynomial whose terms are not all of the same degree, where in determining the degree, one counts the powers of the coefficients as well as the powers of the variable. For example, the formula $ax^2 + bx + c$ is *non-homogeneous* because the degrees of the three terms are respectively 3, 2, and 1. From the modern point of view, one may write $ax^2 + bx + c$ instead of $ax^2 + b^2x + c^3$, which latter expression was required by the traditional geometrical interpretation of the expression as the sums of various cubes.

non-homogeneous This macaronic word should have been *anhomogenic*. See the entry **homogeneous**.

non-secancy This is *the property of not meeting*, and it is said of lines. It, rather than the notion of equidistance, is the basis for the Euclidean definition of parallel lines.

Non sequitur. This is a Latin sentence meaning *It does not follow.* It is used to describe a statement that does not follow from what has gone before.

non-singular See the entries **non-** and **singular**. A matrix M is *non-singular* if its determinant is not 0. The non-singularity of a matrix is equivalent to its having an inverse.

norm The Latin noun *norma*, which is *a carpenter's square for measuring right angles, a standard*, comes from the verb *nosco, noscere, novi, notus*, which means *to become acquainted with*.

normal This is the stem of the Latin adjective *normalis, -e*, which means *made according to the square.*

normalization See the following entry. The practice of forming nouns by adding the Latin suffix *-atio* to verbs themselves formed by adding the Greek suffix *-ize* to adjectives produces macaronic words. There is often an already existing good word, for example, *civility* for *civilization*.

normalize The Latin adjective *normalis* means *pertaining to the* norma *or standard*. This is an example of the bad habit of forming verbs in English by adding the suffix *-ize*, which comes from the Greek infinitive ending -ίζειν, to the stems of Latin adjectives. The word is of the nineteenth century.

Norman A Norman window is a window shaped in the form of a rectangle of base $2r$ and height h surmounted by a semicircle of radius r. It is actually of Roman origin, as can be seen from the Colosseum and other contemporary buildings. It was a form of window resurrected by the Normans and therefore called after them. The Norman window problem asks for the relative dimensions r/h for the maximum area given a fixed perimeter; the answer is *1*. The relative dimensions that maximize the area of the rectangle for fixed perimeter $P = 2h + 2r + \pi r$ are $r/h = 2/(\pi + 2)$.

normed This is the English adjective formed from the noun *norm*. There is no verb *to norm*. See the entry **norm** above.

notation The Latin noun *notatio, notationis* is formed from the verb *noto, notare*, which means *to mark*. The noun *nota*, which means *a mark*, is derived from the fourth principal part of the verb *nosco, noscere, novi, notus*, which means *to get acquainted with*.

nucleus The Latin noun *nux, nucis* means *nut*. *Nucleus* is its diminutive, *a little nut, the nut within the nut, the pit, the kernel*.

null The Latin adjective *nullus, -a, -um* means *not any, no, none*. It is the combination of *ne*, the original Latin particle of negation, and *ullus, any*.

number This is the French *nombre*, which is derived from the Latin *numerus*. (See **numeral** below.) The French alone of nations inserted the *b* into the Latin word.

Numb3rs According to Lynn Arthur Steen ("Distilling Truth from Emotion," a review of the book *Loving and Hating Mathematics, Challenging the Myths of Mathematical Life*, by Reuben Hersh and Vera John-Steiner, in *Science*, July 8, 2011, p. 160), this is the name of a television show. The combination of letters and numbers, or of Latin letters and Greek or Russian letters, such as **numbeя**, to form concoctions attractive to people with no taste is strongly to be deplored. The use of the Cyrillic **я** for the Latin **r** is an example of *illiterate fontese*, the substitution of a foreign letter for a Latin letter that it resembles but does not equal in an attempt to attract the attention of people who are not sufficiently learned to be aggravated by silliness. Also ludicrous is the mixture of capital and lowercase letters; this is the low fashion nowadays, where we have concatenations such as WebWorK, arXiv, and LaTeX.

numeracy This strange word was formed on the analogy of *literacy*. The Romans had the adjective *litteratus, -a, -um*, which meant *branded or marked with* litterae, *letters*, and it was used of slaves maimed in that manner. In later times it acquired the translated meaning *lettered, able to read and write*, and those unable to read and write were the *illiterati*. The nouns *litteratio* and *illiteratio* were also created meaning *literacy* and *illiteracy*, respectively. There is no Latin verb *littero, litterare, litteravi, litteratus*. There is, however, a good Latin verb *numero, numerare, numeravi, numeratus* meaning *to count*. *Numerati* thus means *people who have been counted*, not *people who can count*, and *numeratio* means *counting*, not *ability to count*. The use of English words *numerate* and *numeracy* to mean *able to handle numbers* and *ability to handle numbers* thus contradicts knowledge. A similar even more ugly concoction is the cant word *orality*, which I first found in a faculty course description on October 20, 2011.

numerals, Arabic The Latin word *numeralis, pertaining to number, a numeral*, is first found in the writings of Priscianus (*fl*. A.D. 500). It comes from the noun *numerus, number*, which is related to the Greek νόμος, *law*, and νέμειν, *to distribute*. The Indo-Arabic symbols now current were introduced into the West as a result of the commerce of the Crusades. On account of bad handwriting in the West, the symbols in the left column were transformed over time to the familiar digits in the right column.

$$١ = 1$$
$$٢ = 2$$
$$٣ = 3$$
$$٤ = 4$$
$$٥ = 5$$
$$٦ = 6$$
$$٧ = 7$$
$$٨ = 8$$

$$\tt{9} = 9$$
$$. = 0$$

Before the introduction of the numerals now current, the Jews, Arabs, and Greeks used the letters of their alphabets to denote numbers according to the following system:

1 2 3 4 5 6 7 8 9 10 20 30 40 50 60 70 80

א ב ג ד ה ו ז ח ט י כ ל מ נ ס ע פ

ا ب ج د ه و ز ح ط ى ك ل م ن س ع ف

A B Γ Δ E Ϝ Z H Θ I K Λ M N O Ξ Π

90 100 200 300 400 500 600 700 800 900 1000

צ ק ר ש ת

ص ق ر س ش ت ث خ ذ ض ظ غ

Ϙ P Σ T Y Φ X Ψ Ω ϡ

(Notice that the Greek letters Ϙ, P, Σ, and T do not have the same numerical value as their semitic equivalents ק, ר, ש, and ת.) This system allowed the superstitious to assign numbers to words, each word being given the number equal to the sum of the numerical values of its digits. For example, in Hebrew, the transliteration of *Nero Caesar* is נרון קסר, and the numerical values of the seven digits add up to 666. See the entry **gematria** above.

The Romans used the system of "Roman numerals" too familiar to require an explanation here.

numerator From the Latin noun *numerus, number,* came the verb *numero, numerare, numeravi, numeratus, to count.* The addition of the suffix of agency *-or* to the stem of the fourth principal part produces the noun *numerator, the one who counts.* The use of this word to describe the first of the parts of a fraction is found already in the sixteenth century.

numerology This is a macaronic word, the combination of the Latin stem *numero-* from *numerus,* which means *number,* and the suffix *-ology* routinely used to name subjects with the meaning "study of," derived from the Greek λόγος, *word.* The correct word would have been *arithmology.* As befits a low word, it denominates a low subject, the unreasonable ascription of meaning to various coincidences involving numbers. So, for example, there is no thirteenth floor in hotels, and the pious would interpret 666 as an ominous number in any context. See the entry **gematria**.

O

ob The Latin preposition *ob* is frequently prefixed to verbs. When this happens, its original adverbial sense of *towards, to meet,* is often retained. At other times it is merely pleonastic, just strengthening the sense of the word to which it is appended.

oblate The Latin verb *offero, offerre, obtuli, oblatus* means *to carry or bring (fero) forward, to bring before or against (ob) someone, to offer.* Thus, an *oblation* is an offering, and an *oblate* is a fellow who consecrates himself to the service of a deity. An *oblate spheroid* is an ellipsoid produced by revolving an ellipse around its minor axis. The mathematical sense of the word is due to the idea of the earth being brought forward or made to bulge at the equator by the flattening of the poles.

oblique The Latin adjective *obliquus* means *slanting, sideways*. It is the combination of the pleonastic prefix *ob-* and the adjective *licīnus*, which means *bent*.

oblong This is the stem of the Latin adjective *oblongus* with the meaning *somewhat long, longish*. It is the combination of the pleonastic prefix *ob-* and the adjective *longus*.

obtuse The Latin verb *tundo, tundere, tutudi, tusus* means *to beat, thump, strike repeatedly*. The addition of the pleonastic prefix *ob-* produces the compound verb *obtundo, obtundere, obtudi, obtusus* meaning *to beat upon, to thump, to make dull or blunt by beating*. The English adjective is the stem of the fourth principal part.

octagon See the entry **-gon** above.

octahedron See the entry **-hedron** above.

octant This is the stem of the late Latin *octans, octantis*, formed on the analogy of *quadrant, a fourth part*, from *quadro, quadrare, quadravi, quadratus, to make square*. Thus, an *octant* is an eighth part, although there is no verb *octo, octare*.

-oid The Greek noun εἶδος means *shape*. From it was derived the suffix -ειδής, -ειδές, which was attached to nouns to produce adjectives with the meaning *like a...* Thus, in Euclid we find the adjectives τραπεζοειδής, τραπεζοειδές (shortened to τραπεζωδής, τραπεζωδές) and ῥομβοειδής, ῥομβοειδές (shortened to ῥομβωδής, ῥομβωδές) with the meanings *table-shaped* and *rhombus-shaped*, respectively. The words modified the noun σχῆμα, *shape*, which was often merely understood and not written, so the neuter forms τὸ τραπεζωδές and τὸ ῥομβωδές were used absolutely with the meanings *the table-shaped [figure]* and *the rhombus-shaped [figure]*, respectively, and came into English as nouns. On this analogy the eighteenth-century authors formed nouns *ellipsoid, hyperboloid, paraboloid,* etc., although the solids thereby denominated were not and could not be shaped like the plane figures in the names. *Spheroid,* on the contrary, is the true Greek word σφαιροειδής, and a spheroid actually looks somewhat like the sphere. In modern times this practice of making -*oid* words has gotten out of control, the latest invention being the device called the *android.* The suffix -*oid* comes to us through the mediation of French, where it was more correctly pronounced -*oïde* and not as a diphthong. The *oi* is not in the Greek; it should be either *oï* or *ō.*

-ology See the entry **-logy.**

omega This is the last letter of the Greek alphabet; thus, whereas we say from *A* to *Z,* the Greeks said from *alpha* to *omega.* The capital letter is Ω, and the small letter is ω.

operator The Latin noun *opus, operis* means *work, task.* From the noun was formed the verb *operor, operari, operatus* with the meaning *to work,* and from the stem of its last principal part, by the addition of the suffix -*or,* was formed the noun of agent *operator, he who works.*

opposite The Latin verb *oppono, opponere, opposui, oppositus* means *to put or place (pono) opposite or before (ob).* The English adjective is derived from the fourth principal part.

optimal The Latin adjective *optimus* means *best;* it is the superlative degree of *bonus,* which means *good.* At some point the nominative neuter singular *optimum* was treated like a noun, *the best,* and authors of the nineteenth century superimposed the adjectival suffix -*al* of Latin origin on the stem of what was already an adjective to come up with a technical term meaning *producing or related to the best.*

orbit The Latin noun *orbita, orbitae* means *a wheel-rut,* and, later, *the trajectory of a heavenly body.* It is a formation from the noun *orbis, orbis,* which means *a circle or anything round,* and, more generally, especially in the phrases *urbi et orbi* and *orbis terrarium, the world.*

order The Latin verb *orior, oriri, ortus* means *to rise*. From it was formed the noun *ordo, ordinis* with the meaning *a series, line, row*. An equation of the form $y' + p(x)y = q(x)$ is called a *first-order linear differential equation*; it is first-order because the highest derivative taken is the first.

ordinal The addition of the adjectival suffix *-alis* to the stem of the noun *ordo, ordinis* produced the adjective *ordinalis* with the meaning *pertaining to a series or row*.

ordinary This is the Latin adjective *ordinarius* formed from *ordo* and meaning *according to order, regular, usual, done in the usual manner*. The terminology used in the subject of differential equations is due to Euler (1707–1783). An *ordinary* differential equation is one with no partial derivatives. In the German universities, a *professor ordinarius* was one who was, in American terminology, the holder of an endowed chair.

ordinate From the noun *ordo, ordinis* was formed the verb *ordino, ordinare, ordinavi, ordinatus* with the meaning *to set in order*. See the entry **abscissa**. The *ordinate chords* of an ellipse were the chords perpendicular to and bisected by the major axis. Eventually, the adjective *ordinate* was applied to the top half of these chords only, the half above the major axis.

orient The Latin verb *orior, oriri, ortus* means *to rise*. Its present participle is *oriens, orientis*, which means *rising*. The expression *sol oriens* is *the rising sun*; in this way the *orient* came to mean the *East*. In the morning, a traveller looked for the sun to determine which direction was east, and in that way he got his bearings.

orientation This is the Latin *orientatio, orientationis*, which means *the determination of one's bearings by looking to the morning sun*.

origin The Latin noun *origo, originis*, which means *source, beginning*, is derived from the verb *orior*, which means *to rise*. It is a reference to the sun, from whose location the time of day may be determined.

ortho- The Greek adjective ὀρθός means *straight*. The Latin equivalent is *rectus*. Therefore the prefix *ortho-* should be used with Greek words, and the prefix *recti-* with Latin words. Words like *orthonormal* are therefore low.

orthocenter The Greek adjective ὀρθός means *straight*, and the noun κέντρον means *a point, prickle, spike, spur*. The latter came into Latin as *centrum*.

orthogonal The addition of the Latin suffix *-alis* to a Greek stem is incorrect and has produced this low word. It should have been *orthogonic*, but the damage has been done. If C_1 and C_2 are two families of curves such that whenever a curve in C_1 intersects one in C_2, it does so at right angles, the curves in C_1 are called the *orthogonal trajectories* of the curves in C_2, and *vice versa*. If C_1 is a one-parameter family of curves with general equation $y = f(x:c)$, where c is the parameter, then one finds the general equation of the curves in C_2 by solving the differential equation $y' = -1/(df/dx)$. The theory of orthogonal trajectories was applied by the artist Andrea Pozzo (1642–1709) to solve the *problem of the false dome*: How should one paint a flat ceiling so that it appears from below that there is a dome? His solution is the great *trompe d'oeil* of Sant'Ignazio in Rome, thirty-four meters above the floor. It was necessitated by the fact that Cardinal Ludovisi, who was paying for the construction of the church, died, and the funds needed for the planned dome were no longer forthcoming. A lesser artist tried a few years later to produce the same effect in the chapel of the nearby Palazzo Doria Pamphili, but the ceiling is insufficiently high for the viewer to be fooled.

orthographic The Greek adjective ὀρθός means *straight*, and the adjective γραφικός, *pertaining to writing*, is produced by adding the adjectival suffix -ικός to the stem of the noun γραφή, which means *a drawing*.

orthonormal The Greek adjective ὀρθός means *straight*, and the Latin adjective *normalis* means *pertaining to a carpenter's square (norma), pertaining to any rule or standard*. This is a low word applied to a set of vectors and supposed to mean that any two of them are perpendicular and all are of length *1*.

oscillate The Latin noun *ōs, ōris* means *mouth, face, mask*, and the diminutive *oscillum* means *a little mouth, a little face, a little mask*. The Romans hung *oscilla*, little images of the god Bacchus, from trees, and these were easily swung by the wind. The verb *oscillo, oscillare, oscillavi, oscillatus* thereby came into existence with the meaning *to swing back and forth like the oscilla*.

osculating The Latin noun *ōs, ōris* means *mouth, face, mask*, and the diminutive *osculum* means *a little mouth, a kiss*. From this diminutive was formed the verb *osculor, osculari, osculatus* with the meaning *to kiss*. Struik (p. 419) points out that Johann Bernoulli first used the term *planus osculans*, the origin of our *osculating plane* (*Opera Omnia* IV, pp. 113, 115), thereby giving a name to an object that had already been discussed by the mathematicians of his family and by Leibniz.

osculation The noun *osculatio, osculationis*, which means *kissing*, is formed from the third principal part of the verb *osculor, osculari, osculatus*, which means *to kiss*. See the entry **osculating** above. According to Struik (p. 419), the term is due to Leibniz in his *Meditatio nova de natura anguli contactus et osculi*, Acta Eruditorum, June 1686, pages 289–292.

-ous This is the transformation of the Latin adjectival suffix *-osus*, which is attached to nouns of Latin origin to impart the sense of *full of*.

out-degree The *out-degree* of a directed graph at a vertex is the number of edges leading out of the vertex. See **in-degree**.

oval The Greek noun ᾠόν means *egg*. Related to it is the Latin noun for the same object, *ovum*. The addition of the adjectival suffix *-alis* to the stem of *ovum* produced the word *ovalis* with the meaning *pertaining to an egg*. The *ovals of Cassini* are the plane sections of the torus; the most famous of them is the *lemniscate* of Bernoulli. They were investigated by Giovanni Domenico Cassini (1625–1712), who denied the bulging of the earth at the equators; he was proven wrong by the expedition to Ecuador sponsored by Maupertuis. The ovals may also be produced in the following manner: Let a and b be positive real numbers, and let F_1 and F_2 have coordinates $(-a,0)$ and $(a,0)$, respectively. Then the oval of Cassini is the plane curve consisting of those points P such that the product of the distances from the two fixed points F_1 and F_2 is the constant b^2. The Cartesian equation of the ovals is $(x^2 + y^2 + a^2)^2 - 4a^2x^2 = b^4$. There are three cases, depending on whether $a > b$, $a = b$, or $a < b$. If $a > b$, the graph consists of two little eggs (hence the name oval). If $a = b$, the curve is the lemniscate of Bernoulli. If $a < b$, there is one big peanut-shaped curve.

P

pair The Latin adjective *par* means *equal*. The neuter plural is *paria*. In the development of the French language, the letters r and i were switched, and there was produced the noun *paire*, from which the English word is derived. Such switching is called by the Greeks *metathesis*, and by the Latins *transpositio*.

para- The Greek preposition παρά means *along, beside*, but it can also mean *beyond*. It is now entering the vocabulary of cant: A *paralegal* is someone who performs routine uncomplicated legal chores. *Paratransit certification* is a document of the Commonwealth Public Utilities Commission that allows someone to run a taxi service for Amish people in Pennsylvania.

parabola The parabola is the set of all points in the plane equidistant from a fixed line (the directrix) and a fixed point not on the fixed line (the focus). The distance between the fixed point and the fixed line is the parameter $2p$. The latus rectum of the parabola is $4p$. The Greek noun παραβολή means *a throwing* (βολή) *along* (παρά), *a comparison, an application, a lying along, an equality*. It is derived from the verb παραβάλλω, which means *to throw along*. The explanation of the mathematical term is as follows. Consider the parabola whose equation is $4px = y^2$. Let $P(x,y)$ be a point on the parabola not a vertex, and let S be a square of side $|y|$. Let R be a rectangle whose base is x and whose altitude is the length of the latus rectum of the parabola. Then the area of S is equal to the area of R. The point of intersection of the two parabolas with equations $y = x^2$ and $y^2 = 2x$ is $(2^{1/3}, 4^{1/3})$; the abscissa is therefore the line segment needed to double the cube. The drawings of Leonardo da Vinci illustrated that the parabola is the trajectory of a projectile fired in a vacuum, a fact that was proven by Galileo.

parabolic This is the stem of the Greek adjective παραβολικός. This was one of those cases when the stem of the Latin adjectival ending *-alis* was never superimposed to form the English adjective. We have no word *parabolical* (or *hyperbolical* for that matter), whereas we do indeed say *elliptical*. The parabolic rule of Simpson approximates the area under a positive curve by breaking the curve up into pieces and approximating each piece by a quadratic, that is, parabolic function.

parabolic spiral The parabolic spiral is the polar curve with equation $r^2 = \theta$; it is called *parabolic* because of the similarity of its equation to $y = x^2$. It was first studied by Fermat in 1636. Its design was dug into the mountaintops of Peru by the Incas; for an aerial photograph, see page 55 of the June 1982 issue of *Discover* magazine.

paraboloid This word is formed from the noun *parabola* and the suffix *-oid* on the analogy of the words *rhomboid, spheroid,* and *trapezoid*. If it means anything, it must mean *shaped like a parabola*, but the paraboloid is a solid, and the parabola is a plane figure. The Greeks, when they added the suffix -ειδής, -ειδές to a noun, never promoted it thereby to the next dimension. The church of Sant'Andrea al Quirinale was Bernini's favorite among all his works. Its form is that of an elliptic cylinder surmounted by an elliptic paraboloid. The eccentricity of the elliptical base is big, about $4/5$.

paracompact This is a mongrel word, the union of the Greek preposition παρά, which means *along, beside*, and the Latin participle *compactus*, which means *put together, constructed, built*. See the entry **compact**. There is a section devoted to this concept in Munkres, *Topology: A First Course*, Prentice-Hall, Inc., Englewood Cliffs, New Jersey, 1975. Munkres writes (p. 255), "The concept of paracompactness is one of the most useful generalizations of compactness that has been discovered in recent years." However, the prefix *para-* should not be attached to a noun to indicate a generalization of that noun.

> A space X is *paracompact* if it is Hausdorff and if every open covering $\mathcal{U}$ of X has a locally finite open refinement $\mathcal{B}$ that covers X.

> An *open refinement* of a collection $\mathcal{U}$ of subsets of a topological space X is a collection $\mathcal{B}$ of open subsets of X such that for every element B of $\mathcal{B}$ there is an element A of $\mathcal{U}$ containing B.

> A collection $\mathcal{U}$ of subsets of a topological space X is *locally finite* if every point of X has a neighborhood that intersects only finitely many elements of $\mathcal{U}$.

paradox The Greek adjective παράδοξος means *contrary to* or *beyond* (παρά) *opinion* (δόξα), *contrary to expectation, incredible*. It is a solution to a problem that contradicts the expectation of the superficial observer. For example, the *Petersburg paradox* of Daniel and Nicholas Bernoulli is the most famous paradox in the theory of probability: Peter tosses a fair coin until the first head occurs. If the first head occurs on the ith toss, he gives Paul 2^{i-1} dollars. What does Paul expect to win? Although the game has never been played in which Paul wins more than a few dollars, the mathematical expectation is infinite; this situation was called a *paradox*. The Comte de Buffon inaugurated what has come to be known as the *Monte Carlo method* by attempting to determine Paul's mean winnings by experiment rather than by calculation; he had two of his farmhands at Versailles actually play the game 2,084 times. It cost him $10,057 in prize money, so he was able to conclude that the expectation of the Bernoullis' game was more like 10,057/2,084, or $4.83. The great D'Alembert observed that mean infinity in this game meant that either the axioms of probability or the definition of expectation or both were in need of modification. Emile Borel (1871–1956) noticed that the mean infinity is due to the big prizes for outcomes of tiny probability; if one adds only the first nine terms of the infinite series whereby the mean is calculated, that is, only those terms corresponding to outcomes of probability exceeding .0001, then one would get a result close to that of Buffon, *viz.*, $4.50. In modern times, it is possible to simulate the games that Buffon instructed his serfs to play, for we have tables of random digits. In 10,000 simulated games, Paul's average winnings were $5.22. In 1,000,000,000 simulated games, the student Bryan Hunter found Peter's average winnings to be $13.92. In 2,000,000,000 games, he found the average to be $14.21. In 2002, Brandon Taylor simulated the game 100,000,000,000 times and found that Peter's average winnings were $48.73. Clearly, as the number of simulations goes to infinity, so do Peter's average winnings. The explanation is that the length of time required for sufficiently many tails to occur in succession for Paul to win an astronomical amount of money far exceeds the lifespan of the galaxy, but the definition of expectation takes no notice of this. The mean is infinite *sub specie aeternitatis*.

parallel The Greek adjective παράλληλος means *along* (παρά) *one another* (ἀλλήλων), *side by side*.

parallelepiped This is put together from the Greek παράλληλος, *parallel*, ἐπί, *on*, and πέδον, *the ground*. It is a modern concoction meaning *a solid figure bounded by pairs of parallel planes*.

parallelogram This is the stem of the Greek adjective παραλληλόγραμμος, -ον, which means *bounded by pairs of parallel lines.* (γραμμή means *line.*)

parallelotope This word is compounded from the Greek παράλληλος, *parallel,* and τόπος, *place.* It is a modern concoction generalizing to *n* dimensions what the words *parallelogram* and *parallelepiped* accomplish for two and three dimensions, respectively.

parameter The Greek verb παραμετρέω means *to measure one thing by another, to measure by means of a standard, to compare.* It is the concatenation of the preposition παρά, *along, beside,* and the verb μετρέω, *to measure.* There never was an associated adjective παράμετρος, παράμετρον; the word *parameter* is a modern concoction. The Greeks did have the noun παραμέτρησις and the adjective παραμετρητικός, which meant, respectively, *comparison* and *pertaining to comparison.*

parametric This English adjective was formed as if there were a Greek noun *parameter,* which there is not.

paraproduct This is a modern macaronic concoction "used rather loosely nowadays in the literature to indicate a bilinear operator that, although noncommutative, is somehow better behaved than the usual product of functions." See Árpád Bényi, Diego Maldonado, and Virginia Naibo, "What is a Paraproduct?" *Notices of the AMS,* vol. 57, no. 2, August 2010, pages 858–860, page 858. It is unclear what the authors imagine the force of *para* to be.

parenthesis This is the Greek noun παρένθεσις, which means *a putting* (θέσις) *in* (ἐν) *beside* (παρά). It was transliterated into the Latin *parenthesis,* and the Latin plural, *parentheses,* is the correct English plural.

parsimony The Latin verb *parco, parcere, peperci, parsus* means *to spare.* By the addition of the nominal suffix *-monia* to the fourth principal part was formed the noun *parsimonia* with the meaning *thrift.* The noun ending *-monia* added to a root or verb stem produces a noun denoting an act or the result of an act. See Allen and Greenough, §239. The *principle of parsimony* is the doctrine that that proof is best which relies upon the fewest assumptions and theorems.

parity The Latin adjective *par, paris* means *equal, like, a match*. From it was formed in late times the noun *paritas, paritatis* with the meaning *equality, evenness* (as opposed to being *odd*), whence came the English *parity* through the mediation of French in the usual manner.

part This is the stem of the Latin noun *pars, partis*, which means *a piece or portion of something*.

partial The addition of the adjectival suffix *-alis* to the stem of the noun *pars* produced the late Latin adjective *partialis* with the meaning *pertaining to a part*.

particular The Latin noun *pars, partis* means *a piece, a share*. The addition of the suffix *-cula* produced the diminutive *particula, a small part*. The further addition of the adjectival suffix *-aris* resulted in the adjective *particularis* with the meaning *pertaining to a small part*.

partition The Latin noun *pars, partis* means *a piece, a share*. From it was derived the verb *partio, partire, partivi, partitus* with the meaning *to divide and distribute*. From the fourth principal part came the noun *partitio, partitionis* meaning *sharing*, whose stem is the English word. A partition of a set is a collection of pairwise disjoint subsets of the set whose union is the set.

Peano See the entry **postulate**.

pedal curve The Latin noun *pes, pedis* means *foot*, and from its stem was formed, by the addition of the suffix *-alis*, the adjective *pedalis* with the meaning *a foot long* or *a foot wide*. Let C be a curve, and P any point; P may even be on C. If Q is a point on C, draw the tangent line to C at Q and then the normal line from P to that tangent line. Let the normal line intersect the tangent line at N. Then the locus of N, as Q

varies over C, is called *the pedal curve* of C with respect to P. If the curve C has equation $y = f(x)$ and therefore parametric equations $x = t$, $y = f(t)$, and if we take the fixed point P to be the origin, then the parametric equations of the pedal curve of C with respect to P are

$$x = [t f'(t) - f(t)]f'(t) / [1 + (f'(t))^2]$$

$$y = [t f'(t) - f(t)] / [1 + (f'(t))^2]$$

Pedal curves were first studied by Maclaurin in 1718.

pencil The Latin noun *penis, penis* means *a tail, the membrum virile*. The addition of the suffix *-illus* produces the diminutive *pencillus* with the meaning *a little tail*, whence is derived our noun *pencil*.

pendulum The Latin verb *pendeo, pendere, pependi* means *to hang* (intransitive) and the verb *pendo, pendere, pependi, pensus* means *to hang* (transitive), *to weigh*. The associated adjective *pendulus* means *hanging*, and its neuter form became the English noun. Suppose a point of mass m, suspended from a fixed point by a massless rod of length ℓ, is allowed to swing freely in space under the force of gravity alone. The *pendulum problem* is to determine the angle θ that the pendulum makes from the vertical as a function of time t. The function $\theta(t)$ is the solution of the differential equation $\theta'' + (g \sin \theta)/l = 0$, which cannot be solved in closed form. If, though, one approximates $\sin \theta$ by θ, one finds that for small angles θ, the pendulum swings in nearly simple harmonic motion.

penta- This prefix is derived from the Greek πέντε, which means *five*. In forming new words, it should only be placed before words of Greek origin.

pentagon This is a figure with five angles. See the entries **penta-** and **-gon**.

pentagram This is a figure with five lines. See the entries **penta-** and **gram**. It is especially used of the plane star-shaped figure produced by the five line segments drawn inside a pentagon to make the remaining connections between the vertices.

per This is the Latin preposition meaning *through*. It is regularly attached as a prefix to verbs to intensify the meaning, to give the connotation of thoroughness in the completion of the action. The English use of this preposition in phrases like *per this* and *per that* is cant and to be avoided.

percent This is the abbreviation of the concatenation into one word of the Latin prepositional phrase *per centum*, which means *at the rate of one out of a hundred*.

percentile This is a modern low word created on the analogy of *quartile* by adding the suffix *-ile* to the English noun *percent*.

perfect The Latin verb *facio, facere, feci, factus* means *to do*. The addition of the preposition *per*, *through*, as a prefix imparts the force of thoroughness to the action of the verb and produces the compound *perficio, perficere, perfeci, perfectus* meaning *to do thoroughly, to bring to an end*. The fourth principal part was used as an adjective with the meaning *complete, finished*. The perfect numbers studied by the Hellenes were integers equal to the sum of their proper divisors. In this case, the adjective *perfect* translated the Greek τέλειος. Whether there are infinitely many perfect numbers or just finitely many, and whether there are any odd perfect numbers (for those already known are all even) are open questions. Euler proved that every even perfect number is of the form $2^{n-1}(2^n - 1)$, where $2^n - 1$ is a Mersenne prime. The fact that if $2^n - 1$ is prime, then $2^{n-1}(2^n - 1)$ is perfect is Proposition 36 of Book IX of the Euclid's *Elements of Geometry*.

peri- This prefix is the Greek preposition περί, which means *around*. In forming new words, it should only be placed before words of Greek origin. The corresponding Latin preposition is *circum*.

perigee The Greek adjective περίγειος means *surrounding the earth, around* (περί) *the earth* (γῆ). This is the point in the moon's orbit

240

where it is closest to the earth. The double *e* at the end is the transliteration of the diphthong ει.

perigon As a modern word, this is supposed to mean *an angle* (γωνία) *around* (περί), that is, a 360° angle. It should not be used. There is a Greek word περιγώνιον with the established meaning *a carpenter's set-square*.

perihelion This is the point on the earth's orbit when it is closest to the sun. This point is reached on January 4. It is one of the two points in the orbit where the position and velocity vectors are perpendicular. If r_o and v_o are the distance and speed at perihelion, then the constant rate at which the radius vector from the sun to the earth sweeps out area is $r_o v_o / 2$. The word is composed of the Greek preposition περί, *around*, and the noun ἥλιος, *sun*. It was concocted by Kepler on the analogy of *perigee*.

perimeter The Greek noun περίμετρον means *a measurement* (μέτρον) *around* (περί), *a circumference*.

period The Greek noun περίοδος means *a going* (ὁδός) *around* (περί), *the making a circuit around*.

periodic The addition of the Greek adjectival suffix -ικός to the stem of the noun περίοδος produces the adjective περιοδικός, which means *pertaining to a circuit*.

periphery This is the English form of the stem of the Greek noun περιφέρεια, which means *a carrying* (φέρω = *to carry*) *around* (περί). It was literally translated into Latin by *circumferentia*.

permutation The Latin noun *permutatio, permutationis*, which means *a complete change*, was formed from the fourth principal part of the verb *permuto*. See the following entry. A *permutation* is an ordering of various symbols. The number of different permutations of *n* different symbols taken *r* at a time is $n(n - 1)(n - 2) \cdots (n - r + 1)$.

permute The Latin verb *permuto, permutare, permutavi, permutatus* means *to change (muto) thoroughly (per), to exchange, to interchange*. See the entry **per**.

perpendicular The Latin noun *perpendiculum* means *a plumb-line, a plummet*. The noun is derived from the verb *perpendo, perpendere, perpendi, perpensus*, which means *to weigh (pendo) throroughly (per)*. The prepositional phrase *ad perpendiculum* means *in a straight line*; it is found in Book IV, section 17 of Caesar's *Gallic War*, in the passage describing the construction of the first bridge across the Rhine. The translation is that of William Duncan, *The Commentaries of Caesar*, Tonson and Draper, London, 1753, page 60:

> Tigna bina sesquipedalia paulum ab imo praeacuta, dimensa ad altitudinem fluminis, intervallo pedum duorum inter se iungebat. Haec cum machinationibus immissa in flumen defixerat fistucisque adegerat, non sublicae modo derecte **ad perpendiculum**, sed prone ac fastigate, ut secundum naturam fluminis procumberent....

> Two Beams, each a Foot and a half thick, sharpened a little towards the lower end, and of a length proportioned to the depth of the River, were joined together at a distance of about two Feet. These were sunk into the River by Engines, and afterwards strongly driven with Rammers, **not perpendicularly**, but inclined according to the direction of the Stream....

perpetual The adjectival suffix *-al* is unnecessary, as the Latin *perpetuus* is already an adjective. The superfluous addition of *-al* is quite common in English and French and has produced some of our most common scientific adjectives, such as *mathematical* and *dynamical*. In the former case, the adjective *mathematick* or *mathematic* came slowly to be replaced by the longer form ending in *-al*, whereas in the latter case the form ending in *-al* was invented to indicate a specialized meaning, the earlier shorter form *dynamic* being retained with the other, original, sense. In some cases, like *climactic*, the need was never felt to add the superfluous *-al*. In others, like *philosophic* and *philosophical*, both forms exist without any distinction of meaning. The Latin adjective *perpetuus* originally meant *lasting for a lifetime*, not forever. Thus, Julius Caesar was *Dictator Perpetuus*, dictator for life, not dictator for eternity

perspective The Latin verb *perspicio, perspicere, perspexi, perspectus* means *to see* (*specio, specere, spexi*) *through or throroughly* (*per*). The participle *perspectus* means *ascertained, fully known*. In medieval times, the addition of the adjectival suffix *-ivus* to what was already an adjective produced the low word *perspectivus*; it was used in the phrase *ars perspectiva*, which means *the art of the making of lenses*.

petal This is the stem of the Greek noun πέταλον, *a leaf*, from πετάννυμι, *to spread, stretch out, unfurl*. The corresponding Latin word is *folium*.

phase This word is the English transformation of the Greek noun φάσις, which means *a saying, speech, sentence*. It is related to the verb φημί, which means *to speak*.

phi This is the Greek letter Φ, φ, used in the small case to indicate both the golden ratio and the empty set. I once attended a lecture by a mathematician who used two different forms of writing the lowercase phi, φ and ∅, to indicate two different transformations, calling one "*fee*" and the other "*fie*." This is not recommendable.

philosophy This is the Greek φιλοσοφία, *the love of wisdom*. The mathematician Pythagoras was the first to call himself a *philosopher*, a lover of wisdom. In its original, etymological sense, philosophy is the attempt to give a theoretical explanation of the world and to live according to that explanation. To do this, one has to learn, and that is where mathematics comes in because mathematics comes from μάθεσις, *mathesis*, the Greek word for *learning*, and you cannot explain unless you first know. Mathematics is the condition for the possibility of philosophy. When the Romans translated *mathesis* into Latin, they called it *scientia*, *science*, which means *learning* in their language. Very soon those terms, *mathematics* and *science*, acquired technical meanings; they described those subjects which, *par excellence*, embodied learning and knowledge. The impulse to learn, know, and explain, which is inseparable from an impatience with traditional naïve ideas, the Greeks called θαυμασία, *thaumasia, wonder, amazement, curiosity*; it is this wonder, this amazement, this curiosity, that is the beginning of philosophy. The impatience with traditional naïve ideas may lead to extreme skepticism; the most famous case of this is that of Descartes.

Some of the questions that philosophy asks are:

1) How did this world come about? If it was made by a creator, who was he, and why did he do it?

2) Am I a body, a soul, and a mind, and what happens when I die? What is death, and will some part of me survive it?

3) Did I come into the world a clear slate, or did I come in with some baggage, for example, with some clear and distinct ideas, and if so, where did these ideas come from?

4) What is learning, and how does one go about it?

The Greeks produced many philosophers who gave answers to these questions or declared them unanswerable. The five most important schools of thought were called the five *heresies* because αἵρεσις means *choice* in Greek, and every one chose the school of thought that pleased him best, and followed it, just as American college students choose majors. Of the five great heresies, the first was the School of Plato, called the *Academy*, because it met on the property that legend taught had once belonged to the hero Ἀκάδημος.

In the *Stanza della Segnatura* in the Vatican, Raphael, at the behest of the Pope Julius II, depicted in two famous paintings the heaven of Christianity and the heaven of the Greek philosophers. In the former, called the *Disputa del Sacramento*, the saints, mostly bishops and popes, sit or stand in adoration of Deity, and the books we see are the four Gospels, which are held up by little angels. On the opposite wall, however, Raphael has painted the heaven of the Greek philosophers in the work called *The School of Athens*. Plato and

Aristotle are in the center. Plato points up and carries a book marked *Timeo*, Italian for *Timaeus*, the dialogue in which he describes the creation of the world by a mathematical god. The main activity around him is mathematics; the people there are doing geometry. Indeed, Plutarch (*Moralia, Quaestiones Convivales* VIII, 2, 718c) quoted him as saying θεὸς ἀεὶ γεωμετρίζει, "God is always doing geometry." According to the Platonic point of view, mathematics is the instrument by means of which the Creator works in nature. For Plato, mathematics is the intermediary between the physical world of existence and the mental world of ideas where he felt at home, so he required all students who applied to come under his supervision to study mathematics. If we may believe the twelfth-century Byzantine author John Tzetzes (*Chiliades* viii, line 972), he posted above his door the warning Μηδεὶς ἀγεωμέτρητος εἰσίτω, "Let no one ignorant of geometry enter!" and those who did not qualify he sent away. As a soul and mind, he was restless for the world of ideas; Augustine had a similar feeling when, at the beginning of his *Confessions*, he wrote, *Cor nostrum inquietum est, donec requiescat in te*, "Our heart is restless, until it rests in thee."

The manner in which Euclid's *Elements of Geometry* fits into the philosophical tradition was described by Bertrand Russell (1872–1970) in a wonderful paragraph:

> The influence of geometry upon philosophy and scientific method has been profound. Geometry, as established by the Greeks, starts with axioms that are (or are deemed to be) self-evident, and proceeds, by deductive reasoning, to arrive at theorems that are far from self-evident. The axioms and theorems are held to be true of actual space, which is something given in experience. It thus appeared to be possible to discover things about the actual world by first noticing what is self-evident and then using deduction. This view influenced Plato and Kant, and most of the intermediate philosophers. When the Declaration of Independence says "we hold these truths to be self-evident," it is modeling itself on Euclid. The eighteenth-century doctrine of natural rights is a search for Euclidean axioms in politics. ("Self-evident" was substituted by Franklin for Jefferson's "sacred and undeniable.") The form of Newton's *Principia*, in spite of its admittedly empirical material, is entirely dominated by Euclid. Theology, in its exact scholastic forms,

245

takes its style from the same source. Personal religion is derived from ecstasy, theology from mathematics, and both are to be found in Pythagoras. (Bertrand Russell, *History of Western Philosophy and Its Connection with Political and Social Circumstances from the Earliest Times to the Present Day*, Folio Society, London, 2004, pp. 36–37)

Plato, in the dialogue that Raphael painted him holding in the *School of Athens*, and which has had more influence than anything else that he wrote, described the creation of the world by the mathematical god in conformity with the laws of plane and solid geometry. The regular, "Platonic," solids, upon which he founded the chemistry of the four elements, became the subject of the thirteenth and final book of Euclid's *Elements of Geometry*, to which the preceding twelve books were but the prerequisite. All the Platonic philosophers studied, and most, like Proclus (410–485) and Simplicius (sixth century A.D.), wrote, commentaries upon Euclid. When Heiberg published the critical edition of the Greek Euclid for Teubner (1883–1888), it was agreed that the commentary by al-Nayrizi (النيريزي, *fl.* A.D. 900), was so important (in part because it preserves, in Arabic translation, the commentaries of previous Platonic philosophers that are lost in the original), that the medieval Latin version of it by Gerard of Cremona (1114–1187) was published as a sixth, supplementary, volume in 1899.

When the Arabs emerged from their peninsula in the second quarter of the seventh century, they possessed neither philosophy nor mathematics, as far as the world noticed. The first expression of philosophy in Islam (*circa* 757) was المعتزلة, the *Mutazilite* ("schismatic") school, which developed as a result of the introduction of the deductive method of reasoning from first principles that the Arabs had learned from Greek mathematics and logic; this development was possible in part because of the translations of Euclid being made at the time from the Greek into the Arabic language. We may recall in this regard the complaint of Eusebius against some contemporary heretics, quoted by Gibbon near the end of the fifteenth chapter of the *History of the Decline and Fall of the Roman Empire*:

They presume to alter the holy scriptures, to abandon the ancient rule of faith, and to form their opinions according to the subtle precepts of logic. The science of the Church is neglected for the study of geometry, and they lose sight of Heaven while they are employed in measuring the earth. Euclid is perpetually in their hands....

The *Mutazilites* insisted on the allegorical interpretation of the Quran where that book seemed to contradict science and condemned as fatal to the exercise of human reason the absolute predestination of all events by the Deity as taught by السّنّيون, the *Sunni* Muslims, the people who follow habitual practices sanctioned by السنة, the *Sunna*, that is, *tradition*. The translation and study of Euclid was a main activity among the learned in the early Abbasid society of eighth- to tenth-century Baghdad; one can no longer say, however, that the translation of Greek mathematical texts was a major enterprise of the بيت الحكمة , *the House of Wisdom*, established by the *Mutazilite* Caliph al-Mamum (813–833) (Sonja Brentjes, Review of "Le Développement de la Géometrie aux IXe–XIe Siècles, Abu Sahl al-Quhi," *Historia Mathematica* 34 [2007], p. 347). *Mutazilite* doctrines linger on today in the other great heretical community of Islam, الشّيعيون, the *Shiite* Muslims, the people of الشيعة, *al-shia*, that is, *the faction*, of Ali. The first Muslim philosopher, الكندي, al-Kindi (born 803), studied the work of Euclid and then taught the Platonic doctrine that no one could be a philosopher without first being a mathematician. The orthodox Muslims considered the philosophers heretics, and there was a reaction in the ninth century. This led to the rise of the compromise movement of المتكلمون, the *Mutakallimites* ("logicians"), who attempted to reconcile traditional beliefs with the rational methods of the philosopher-mathematicians. Al-Nayrizi was a contemporary of this movement and was influenced by it, as was الفارابي, al-Farabi (*circa* 870–950), a student of whom established in Baghdad the association called اخوان الصفاء, the *Brethren of Sincerity*, where the study of Euclid was continued. The reconciliation of Islamic faith and reason, in so far as possible, was the goal of ابن سينا, Avicenna (980–1037), who insisted on precise definitions of all technical terms in the manner of the mathematicians. Much of what is found in Albertus Magnus (1193–1280) goes back to Avicenna, whom Albert considered the apex of medieval Islamic thought.

pi This is the name of the Greek letter Π, π corresponding to our letter *P, p*. The capital letter is used to indicate a *product*. It was a theorem of Archimedes (*On the Measurement of Circles*) that the ratio of the circumference to the diameter is the same for all circles. Struik (p. 347) has traced back the use of the small case letter π to indicate this ratio to William Jones, *Synopsis palmariorum matheseos*, London, 1706, page 243. (See David Eugene Smith, *History of Mathematics*, vol. II, p. 312.) The notation was adopted by Euler as early as 1736 (Struik, p. 347) and is to be found in his *Introductio in Analysin Infinitorum*, Lausanne, 1748, chapter viii. The fact that a line of length π cannot be constructed with unmarked straightedge and compass was proved by Lindemann in 1881. This has not deterred many people from claiming to have done so; they are cruelly denominated *circle-squarers*. The fact that π is not constructible means that it is not rational; the fact that it is transcendental was established by Lindemann in the following year 1882.

piecewise Weekley says that the word *piece* is of Celtic origin and that the Celtic original entered medieval Latin as *petia, -ae*. The original meaning was that of "a fixed amount or measure." It then proceeded into French, from which it entered English. For the suffix *-wise*, see **clockwise**.

piercing point Weekley derives the verb *pierce* from the Latin verb *pereo, perire, perivi, peritus*, which means *to go (ire) through (per)*. See the entry **point**.

piriform The Latin noun *pirus* means *pear-tree*, and *pirum* is *a single pear*. *Forma* means *shape*. There is no evidence of any adjective *piriformis* in literature, but the construction is a good one in accordance with scientific principles. According to Lawrence (pp. 148–150), the word *prirform* was first used by De Longchamps in 1886 for the plane curve of that name, whose Cartesian equation is $a^4y^2 = b^2x^3(2a - x)$, where *a* and *b* are non-zero constants.

place The Greek adjective πλατύς, πλατεῖα, πλατύν means *wide, flat*; the adjective from the phrase πλατεῖα ὁδός, *wide street* or *broadway*, came into Latin as *platea* with the meaning *a street*. This is the origin of the French and English *place*, of the Italian *piazza*, and of the Spanish *plaza*.

planar The addition of the adjectival suffix *-aris* to the stem of what was already an adjective, *planus*, produced the late Latin superadjective *planaris* with the meaning *on a level surface, flat, plane*. Finkbeiner and Lindstrom (*A Primer of Discrete Mathematics*, W. H. Freeman and Company, New York, 1987, p. 230) define a graph *G* to be *planar* if and only if it is isomorphic to a graph drawn in the plane with no pair of edges crossing.

plane The Latin adjective *planus* means *even, level, flat*. This word came into English as *plain*. In the sixteenth century, the spelling variation *plane* came into use. Before the *Dictionary* of Dr. Johnson (1755), English spelling was like Greek grammar; you could do almost anything you liked.

plus This word is the Latin comparative adjective and adverb meaning *more*. The positive degree of the adjective is *multus, much*, and its superlative degree is *plurimus, most*. The positive and superlative degrees of the adverb are *multum* and *plurime*.

point This word is the transfiguration of the Latin noun *punctum*, a prick, from *pungo, pungere, pupugi, punctus, to stab*. In mathematics, *punctum* and *point* translate the Greek σημεῖον, *a mark*, of Euclid. The *problem of points* in the theory of probability arose from a seventeenth-century dispute that was brought to the attention of Pascal in an attempt at mediation. Peter and Paul toss a fair coin. If it comes up heads, Peter gets one point; if it comes up tails, Paul gets one point. The winner is the first fellow to win *n* points, and the prize is 1,000 francs. If the game is interrupted when Peter has *i* points and Paul has *j* points, and it cannot be resumed, how should the prize money be divided between the two of them?

polar The addition of the adjectival suffix *-aris* to the stem of the Latin noun *polus* produced the adjective *polaris* with the meaning *pertaining to a pole*. The *polar coordinate system* of Jakob Bernoulli (1654–1705) consists of a directed ray, called the *initial line*, emanating from a fixed point *O*, called the *pole*. The line segment connecting the point *P* to *O* is called the *radius vector* to *P*. To obtain a pair of polar coordinates for *P* (for they are not unique), one rotates the initial line until it or its extension in either direction is aligned with the radius vector; the angle of rotation θ is the second polar coordinate. By universal agreement, the angle obtained by counterclockwise rotation is considered positive, whereas one obtained by rotating clockwise is counted negative. The first polar coordinate *r* is the directed distance from *O* to *P*; it is counted positive if the positive part of the initial line is aligned with the radius vector and negative otherwise. The importance of the polar coordinate system is due to Kepler's second law since in this system it is easy to calculate the area swept out by the rotating radius vector.

pole The Greek noun πόλος, *the end of an axis*, came into Latin as *polus* and from thence entered French and English as *pole*. Knopp defines a *pole* in complex function theory as follows:

> According as this entire function is an entire transcendental or an entire rational function, i.e., according as that part of the [Laurent] expansion involving the descending powers of z-z_0 contains an infinite number or only a finite number of terms (but then at least one), z_0 is called an *essential* or a *non-essential singularity*. In the latter case, z_0 is also called briefly **a pole**. (Knopp, Part I, p. 123)

poly- The Greek adjective πολύς, πολλή, πολύ means *much, many*. The stem πολυ- is transliterated into English by *poly-*, the letter *y* being used to express the *upsilon*, which was evidently pronounced like *i* or the German *ü*. This stem was used as a prefix to indicate *much* or *many*.

polyadic This word is formed on the analogy of *dyadic* and *triadic*. It was as if there were a Greek noun πολυάς, πολυάδος with the meaning *the number "muchness."*

polygon This is the stem of the Greek adjective πολύγωνος, πολύγωνον, *having many angles*. See the entries **poly-** and **-gon**.

polyhedron This is the stem of the Greek adjective πολύεδρος, πολύεδρον, *having many seats*. A *polyhedron* is a solid bounded by plane polygons. See the entries **poly-** and **-hedron**.

polymath project This is supposed to mean a project in which a large group, even hundreds, of mathematicians work together. According to Krantz (*Notices of the AMS*, August 2011, p. 893), it was initiated by Timothy Gowers. However, the word *polymath* is used incorrectly here. *Polymath* is an established noun, not an adjective; a *polymath* is someone who has universal knowledge. *Poly* here refers to the knowledge, not to the number of people who have it.

polynomial This is a low word with many faults. It was invented in the seventeenth century on the analogy of *binomial*, itself a low word. It has the Greek prefix πολύ- meaning *many*, the stem of the noun ὄνομα, meaning *name*, minus the first *omicron*, the extra letter *-i-* inserted, and then the stem of the Latin adjectival suffix *-alis*. The correct word to get the idea across would have been either *polyonomic* or *multinominal*.

polytope This word is compounded of the adjective πολύ, which means *much, many*, and the noun τόπος, *place*. An *n*-dimensional polytope is a compact subset of R^n whose interior is connected, which is the closure of its interior and whose boundary consists of hyperplanes. The word does not suggest what it names.

pons asinorum This Latin phrase means *the bridge of asses*. It is a name given to the fourth Proposition of the first Book of Euclid's *Elements*, the idea being that a weak student would manage to plow through the first three propositions but would get stuck at the fourth.

porism The Greek verb πορίζω means *to bring, conduct, fetch, convey*, and from it is derived the noun πόρισμα with the meaning *a procuring, a means of acquiring, profit, gain*. It became the mathematical technical word for a result that follows immediately, as a sort of extra, from another result. In English we commonly use the word *corollary* for such a proposition.

position The Latin verb *pono, ponere, posui, positus* means *to put*. From its fourth principal part is derived the noun *positio, positionis, a putting or placing*, whose stem is the English noun.

positive The Latin verb *pono, ponere, posui, positus* means *to put or place*. The addition of the adjectival suffix *-ivus* to the stem of what was already an adjective, the past participle *positus*, produced in the Middle Ages the superadjective *positivus*. The modern technical sense of *positive* and *negative* is traceable at least as far back as 1704.

postulate The Latin verb *postulo, postulare, postulavi, postulatus* means *to request, claim, demand*. It is the Latin translation of the Greek verb αἰτέω, *to ask for*. Thus, Boëthius translated the corresponding Euclidean technical term αἰτήματα by *postulata* (neuter plural); the postulates are the five geometrical statements required to be accepted before proceeding in the investigation of the *Elements of Geometry*. The Latin verb is related to another verb *posco, poscere, poposci,* which means *to ask earnestly*.

A very good junior mathematics major once spoke to me at dinner of *Peano's postulates*, but he pronounced the Italian's name as if it were *Peeno*. I thought that either he was joking or he was making the typical American mistake of pronouncing the name as if it were English, but he told me, *No*, that he had actually learned that pronunciation from his professor, a young Ph.D. in mathematics. It may therefore be of no little use to discuss the topic of the correct pronunciation of names.

Since everyone is entitled to his own name, it seems reasonable to hold that a fellow's name should be pronounced as he himself pronounces it. This is usually in accordance with the laws of the language that he speaks. Also reasonable is to grant exceptions to this rule in the case of immemorial custom or when the sounds in question do not exist in English.

We have noticed before the phenomenon that each nation is naturally disposed to pronounce foreign names in its own manner. No one insists that we pronounce *Caesar* or *Cicero* in the Latin manner, for that would result in their not being understood. Since the majority of the decent literature in the world is not written in English, the question as to whether the English pronunciation of foreign words is right or wrong is of some interest. While a student at Kenyon College in Gambier, Ohio, *anno* 1965, I was told by the Shakespearean scholar Patrick Cruttwell that it was all right to say *Don Jewan* and *Don Kwixote*. If such is the universal practice in England, then so be it, and let us by all means say *Don Jewan* and *Don Kwixote* in that country, but in America, the Spanish pronunciation is the only one that can be tolerated because it is traditional; the native pronunciation prevails over the English here except among the most ignorant speakers. Americans must therefore continue to speak of Don Juan and Don Quixote pronounced in the Spanish manner.

English pronunciations of foreign names, even though they contradict knowledge, eventually prevail because there is strength in numbers, and there will always be more unlearned then learned people, and the error becomes the usage. This development promotes ignorance, and it should always be opposed if the case is not yet closed, even if we are doomed to fail, because the struggle is beautiful and there is always some small hope of prevailing.

potential The present participle of the verb *possum, posse, potui, to be able*, is *potens, potentis*. From it proceeds the noun *potentia, power*. The addition of the adjectival suffix *-alis* to its stem produced the word *potentialis* meaning *pertaining to someone who has power*.

pound The Latin verb *pendo, pendere, pependi, pensus* means *to cause to hang down, to weigh, to pay*. From this verb was derived the noun *pondus, ponderis* with the meaning *weight*. The English noun *pound* is the corruption of *pondus*.

pound symbol # The traditional abbreviation for *pound* is *lb.*, the abbreviation of the Latin noun *libra*, which means *a balance, a scale*, that by means of which money was paid out. The abbreviation *lb.* was abbreviated even more in the Middle Ages to £, a symbol that has survived in Britain. The use of the symbol # for *pound* promotes confusion since it is already a symbol for *number*.

power The Latin adjective *potis* means *able, capable*, and *sum, esse, fui, futurus* is the verb *to be*. The verb *possum, posse, potui* is composed of the parts *potis* and *esse* and means *to be able*. In the development of French, *posse* became *pouoir*, and then a middle *v* was inserted because of the influence of the similarly sounding verb *avoir* and *devoir*, where the *v* actually represented the Latin letter *b* (*habere, debere*). Thus *posse* became *pouvoir*. The *v* then became a *w* in English, and we now have *power*. Such is the explanation of Weekley.

PowerPoint This is an example of that childish practice, which appeals to the intellectually limited, wherein words are run together, often, as in the case of *LaTeX* and *WeBWorK*, with capitals and lowercase letters mingled arbitrarily. I have just received a desk copy of the seventh edition of Stewart's *Calculus*, which comes with a DVD entitled *PowerLecture with JoinIn and ExamView*. Another example of this nonsense is *WebAssign*, which is advertised as a "Homework System for Calculus." The only effective homework system is for the instructor to collect and examine the homework, but this is not what the common sort of teachers see as their duty, and as a result most students do not have the benefit of competent feedback on their written work, except when that work is in the form of tests and the ever more frequent quizzes.

pre- This is the English suffix derived from the Latin preposition *prae*, which means *before* in the temporal sense.

precalculus This is modern cant for studies that should come before calculus is attempted. It is cataloguese. The combination of precalculus and calculus in one remedial calculus course was a fad attempted without success in recent decades.

predicate The Latin verb *dico, dicere, dixi, dictus* means *to say*; related to it is the verb *dico, dicare, dicavi, dicatus* meaning *to consecrate, devote to the gods*. From the second verb proceeds the compound verb *praedico, praedicare, praedicavi, praedicatus* meaning *to make publically known, to proclaim, to preach, to assert, to praise*. The noun *predicate* comes from the fourth principal part.

premise The Latin verb *praemitto, praemittere, praemisi, praemissus* means *to send (mitto) before (prae), to send on ahead*. The adjective *praemissus* means *aforementioned* and the feminine *praemissa* was used independently in the Middle Ages with the noun *propositio* understood to mean *the previously mentioned [proposition]*. The extra *s* was dropped by some later careless writers, who added a final *e*; the spelling *premiss* is more accurate.

preperiodic If *f* is a mapping from the complex plane into itself, a point z is *preperiodic* if the orbit of z under *f* is eventually periodic. The term is macaronic and low, the combination of the Latin preposition *prae* and the Greek adjective *periodic*.

primary The Latin adjective *primus* means *first*. The addition of the suffix *-arius* to the stem produced *primarius* with the meaning *belonging to the first rank*. Similarly, *secondarius* means *belonging to the second rank*, *tertiarius* means *belonging to the third rank*, etc. These words came into English as *primary, secondary, tertiary*, etc.

prime The Latin adjective *primus* means *first*. It is actually the superlative degree of the adjective *priscus*, which means *ancient*. The *numeri primi* first became *primi* and then *primes* in English. The *Mersenne primes* are prime numbers of the form $2^n - 1$. (Not all numbers of the form $2^n - 1$ are prime.) At the moment, according to the eponymous

Wikipedia article, there are forty-seven known Mersenne primes. There is a one-to-one correspondence between Mersenne primes and even perfect numbers. *Fermat primes* are prime numbers of the form $2^N + 1$, where $N = 2^n$ for $n = 0, 1, 2,...$ (Not all numbers of the form $2^N + 1$ are prime.) There are currently five known Fermat primes: 3, 5, 17, 257, and 65,537. (The eponymous Wikipedia article tells us that $2^N + 1$ is not prime for $2^5 \leq N \leq 2^{32}$.) The Fermat primes are important because a regular *n*-gon is constructible with unmarked straightedge and compass if and only if *n* is a power of 2 greater than or equal to 4, or if $n = 2^m p_1 p_2 p_3 \cdots p_k$, where p_1, p_2, $p_3,...,p_k$ are different Fermat primes, $m \geq 0$ and $k \geq 1$. The Greek word for *prime* used by Euclid is πρῶτος ἀριθμός, literally *a first number*.

primitive The Latin noun *primitiae* means *the first fruits*. The addition of the suffix *-ivus* to the stem produced the adjective *primitivus* with the meaning *pertaining to the first fruits*.

principal The addition of the adjective suffix *-alis* to the stem of *princeps, principis* produces the adjective *principalis* with the meaning *pertaining to the first rank*. For *princeps*, see the entry **principle**.

Principia Mathematica This is the title of the book of Russell and Whitehead. Newton's masterpiece was entitled *Philosophiae Naturalis Principia Mathematica*. See the following entry.

principle The Latin noun *principium* means *beginning*. It is derived from the adjective *princeps*, which means *foremost* and is composed of the elements *primus* (*first*) and *capio* (*to take*). The *princeps* is someone who takes the first place; from it we get our noun *prince*.

prior This Latin word is the comparative degree of the adjective *priscus, ancient*. *Prior* probabilities are those that are accepted prior to experiment. *Posterior* probabilities are those probabilities assigned after the occurrence of the experiment. The manner of adjusting the prior probabilities to obtain the posterior ones is given by Bayes' theorem.

prism The Greek verb πρίζω (also πρίω) means *to saw*. From it was derived the noun πρίσμα, πρίσματος, *a piece sawed off*. A *prism* is

> a polyhedron with two congruent and parallel faces, called the *bases*, and whose other faces, called *lateral faces*, are parallelograms formed by joining corresponding vertices of the bases; the intersections of lateral faces are called *lateral edges*. (James and James, *Mathematics Dictionary*, Students edition, D. Van Nostrand Company, Inc., Princeton, New Jersey, 1964, p. 306)

prismatoid This word is the adaptation of the Greek adjective πρισματοειδής, *like a prism*, and is composed of the stem of the Greek noun πρίσμα, πρίσματος, *a piece sawed off*, and the noun εἶδος, *shape*. See the entry **prism**. A *prismatoid* is

> a polyhedron whose vertices all lie in one or the other of two parallel planes. (James and James, *Mathematics Dictionary*, Students edition, D. Van Nostrand Company, Inc., Princeton, New Jersey, 1964, p. 306)

prismoid This word is composed of the Greek noun πρίσμα, which means *a sawed off piece*, and the noun εἶδος, *shape*. It is a modern word, in use in English by the beginning of the eighteenth century. A *prismoid* is

> a prismatoid whose bases are polygons having the same number of sides, the other faces being trapezoids or parallelograms. (James and James, *Mathematics Dictionary*, Students edition, D. Van Nostrand Company, Inc., Princeton, New Jersey, 1964, p. 306)

The words *prismatoid* and *prismoid* are formed from the same Greek elements; the former is formed correctly and the latter incorrectly. One speaks of *mathematics*, not *mathemics*.

pro- In Greek, the preposition πρό means *before*, both spatially and temporally. In Latin, however, the preposition *pro* means *for* in the sense of *in favor of, on behalf of*, as in the phrase *pro bono*, *for the public good*, or *before* in the spatial sense, *in front of*. *Before* in the temporal sense in Latin is *ante*. To use the prefix *pro-* with the Greek, temporal meaning before a word not of Greek origin results in an ugly, low concoction like *proactive*, imagined by the multitude to be the opposite of *reactive*.

probability The Latin adjective *probus* means *good, excellent, fine*. From it is derived the verb *probo, probare, probavi, probatus* meaning *to make or find good*. The addition of the adjectival suffix *-abilis* to the stem of the adjective produces *probabilis*, which means *that which can be shown to be good*.

> A wise man, therefore, proportions his belief to the evidence. In such conclusions that are founded on an infallible experience, he expects the event with the last degree of assurance, and regards his past experience as a full *proof* of that event. In other cases, he proceeds with more caution: He weighs the opposite experiments: He considers which side is supported by the greatest number of experiments: To that side he inclines, with doubt and hesitation; and when at last he fixes his judgment, the evidence exceeds not what we properly call probability. (David Hume, *An Enquiry concerning Human Understanding*, Chapter X, "Of Miracles," p. 344, in *Essays and Treatises on Several Subjects*, a new edition, A. Millar in the Strand, and A. Kincaid and A. Donaldson, at Edinburgh, London, 1758)

Laplace's definition of probability was that if an experiment has n equally likely outcomes, and if S is a collection of m distinct outcomes, then the probability that the experiment will result in one of the outcomes in S is m/n. It was one of the major problems of mathematics to determine how to extend this definition to the case when the outcomes are not equiprobable or when there are infinitely many of them. It was a theorem of Vitali that it is not possible to define a probability distribution over a continuum so that a probability can be assigned to every set of outcomes. Those subsets to which no probability can be assigned are called *non-measurable* sets.

problem The Greek verb βάλλω means *to throw*, and the prefix πρό imparts the adverbial force of *forward*. The result of their combination is the verb προβάλλω, which means *to throw forward, to put forward, nominate, propose*. The associated noun πρόβλημα (plural

258

προβλήματα) was the Greek name for those propositions of Euclid that were constructions, for example, Propositions 1, 2, and 3 of Book I; the others were called θεωρήματα, *theorems*.

produce The Latin verb *produco, producere, produxi, productus* means *to lead* (*duco*) or *bring forward* (*pro*), *to extend*.

product This word is the stem of the fourth principal part *productus, -a, -um* of the verb *produco*. See the previous entry.

program The Greek noun πρόγραμμα means a *public proclamation, notice, injunction, advice*. It is associated with the verb προγράφω, which means *to write* (γράφω) *beforehand* (προ-).

progression This word is the stem of the Latin noun *progressio, progressionis*, which means *an advance*. It is derived from the third principal part of the verb *progredior, progredi, progressus*, which means *to go* (*gradior*) *forward* (*pro*).

project The Latin verb *proicio, proicere, proieci, proiectus* means *to throw* (*iacio*) *forward* (*pro*).

projection See the previous entry. The English word is the stem of the Latin noun *proiectio, proiectionis*, which means *a throwing forward*, formed from the fourth principal part of the verb *proicio*.

projective See the entry **project** above. This English adjective of the seventeenth century was formed by adding the stem of the Latin adjectival suffix *-ivus* to the stem of the participle *proiectus, -a, -um* of the verb *proicio*; the neuter singular *proiectum* was evidently viewed as a noun, *a thing that jutted forward from a building*. The adjective *projective* thus means *having to do with the projection of some figure onto some surface*.

prolate The Latin verb *profero, proferre, protuli, prolatus* means *to carry* (*fero*) or *bring forward* (*pro-*), *to lengthen*. The *prolate ellipsoid* is the quadric surface produced by revolving an ellipse around its major axis, the axis that has been lengthened with respect to the other, minor, axis.

The *prolate trochoid* is the trochoid produced by extending the distance from the fixed point to the center of the rolling circle beyond the circumference of the rolling circle. See the entry **trochoid**.

proof This is the corruption of the late Latin noun *proba*, which means *a demonstration, a trial*. It is derived from the adjective *probus*, which means *good, excellent, fine*. From it is derived the verb *probo, probare, probavi, probatus* meaning *to make or find good*. When this word came into Italian as *provo, provare*, it had already acquired the additional meaning *to test, to try*. See the excerpt from Hume in the entry **probability**.

proper The Latin adjective *proprius* means *one's own*.

property The Latin adjective *proprius* means *one's own*. From it was formed the noun *proprietas, proprietatis* with the meaning *that which is one's own*. *Property* is the metamorphosis of *proprietas*.

proportion This is the amalgamation into one word of the Latin prepositional phrase *pro portione*, which means *with respect to one's share*. The phrase then became the late noun *proportio, proportionis*.

proportional The suffix *-alis* was added to the stem of the late Latin noun *proportio, proportionis* to produce the adjective *proportionalis* with the meaning *pertaining to a proportion*.

proposition The Latin word *propositio, propositionis* means *a setting (positio) before (pro-)*. It is derived from the fourth principal part of the verb *propono, proponere, proposui, propositus*, which means *to put before*.

prosthaphairesis This word is the Greek προσθαφαίρεσις, a combination of πρόσθεν (*in front, additionally*) and ἀφαίρεσις (*taking away, subtraction*). It means *adding and subtracting* in Greek. It is used of the formulas for *sin (A + B)*, *cos (A + B)*, etc. The formulas are due to Ptolemy of Alexandria (second century A.D.) and are essentially to be found in his *Almagest*. The spelling *prostaphairesis* is wrong since the Greek word has *theta*, not *tau*.

protasis This is the Greek noun πρότασις, which means *a stretching forward, that which is put forward, a premise, the antecedent clause in a conditional sentence*. It is derived from the verb τείνω, *to stretch*, and the prefix πρo-, *forward, before*.

protractor The Latin verb *protraho, protrahere, protraxi, protractus* means *to drag (traho) forward (pro-)*. The noun of agent *protractor*, formed by adding the suffix *-or* to the stem of the fourth principal part, is *the fellow who drags forward*.

prove The Latin adjective *probus* means *good, excellent, fine*. From it is derived the verb *probo, probare, probavi, probatus* meaning *to make or find good*.

pseudo- This prefix comes from the Greek adjective ψευδής, which means *false*. To attach it to a word that is not of Greek origin is low style, though common. Thus words like *pseudoidentity, pseudoperfect, pseudoprime, pseudotangent, pseudovertex* are all low. The proper suffix to have added would have been *falsi-* from *fallo, to lie*, but it is too late, and the damage has been done. The use of this prefix is abused by those who employ it without the necessary discrimination.

pseudoequicontinuity This noun is an example of a superfluity of prefixes. It would be better written *pseudo-equicontinuity*; the hyphen should be written to circumvent the momentary hesitation in reading caused by wondering whether the *oe* is a diphthong. Royden (p. 177) defines a family $\mathscr{F}$ of functions from a topological space X to a metric space $\{Y, \sigma\}$ to be *equicontinuous* at the point $x \in X$ if given $\varepsilon > 0$ there is an open set O containing x such that $\sigma(f(x),f(y)) < \varepsilon$ for all y in O and all $f \in \mathscr{F}$. A sequence $\{f_n\}$ of functions from a compact metric space X into a Banach space Y is *pseudo-equicontinuous* if, for every $\varepsilon > 0$ and each x in X, there is a positive integer n_0 and a neighborhood U_x of x such that $\| f_n (x) - f_n (y)\| < \varepsilon$ if $n \geq n_0$ and $y \in U_x$. (See Piccinini, Stampacchia, and Vidossich, *Ordinary*

Differential Equations in Rⁿ: Problems and Methods, Springer-Verlag, New York, 1984, p. 140.)

pseudoidentity matrix *Pseudo* comes from the Greek word ψευδής, *false*, and *identity* is derived through French (*identité*) from a fifth-century A.D. Latin noun *identitas, identitatis* meaning *sameness*, from *idem*, the same. The macaronic combination is a sign that the word is low. It would be preferable to write *pseudo-identity*; the hyphen would circumvent the momentary hesitation in reading caused by wondering whether the *oi* is a diphthong. A *pseudo-identity matrix* is a matrix that would become an identity matrix if certain columns or rows of zeroes were removed.

pseudometric This word is the combination of the Greek prefix ψευδο-, *false*, and the stem of the Greek adjective μετρικός, *pertaining to measurement*. Royden (p. 129) defines a *pseudometric space* as a pair $\{X, \varrho\}$ such that ϱ satisfies all the conditions of a metric except that $\varrho(x,y) = 0$ need not imply that $x = y$.

pseudonorm This word is the macaronic combination of the Greek prefix ψευδο-, *false*, and the stem of the Latin noun *norma*, which means *a carpenter's square for measuring right angles, a standard*. Royden (p. 183) defines a *pseudonorm* to be a non-negative real-valued function $\| \, \|$ defined on a vector space X such that

$$\|x + y\| \leq \|x\| + \|y\| \text{ and } \|ax\| \leq |a| \|x\|.$$

Thus, other vectors besides the zero vector may have length *0*.

pseudoperfect This low English word has been constructed from the Greek prefix *pseudo-* and the adjective of Latin origin *perfect*. A *pseudoperfect number* is a positive integer that is the sum of some of its proper divisors. Thus *12* is *pseudoperfect* because *12 = 1 + 2 + 3 + 6*.

pseudoprime This low English word has been constructed from the Greek prefix *pseudo-* and the adjective of Latin origin *prime*. The name *pseudoprime* is granted by some people who should know better to a number that is not a prime but that satisfies one or more, but not all, of the hypotheses of a theorem that gives a sufficient condition for a number to be a prime. It is hard to think of a positive integer that does not fall into this category. See the entries **pseudo-** and **prime.**

pseudosphere This modern word is correctly compounded of the prefix ψευδο-, *false,* and the noun σφαῖρα, *a sphere.* The *pseudosphere* is the surface of revolution produced by revolving a tractrix about its asymptote. It acquired the name because it is a surface of constant *negative* curvature.

pseudotangent This very low English noun has been constructed macaronically from the Greek prefix *pseudo-* and the Latin adjective *tangent.* A path is a pseudotangent to a curve at a point if it passes through the point only once.

pseudovertex This low English word has been constructed from the Greek prefix *pseudo-* and the Latin noun *vertex.* The points where the minor axis of an ellipse intersects the ellipse are called the *pseudovertices* of the ellipse.

Ptolemaic system See the entry **epicycle.**

pure The Latin adjective *purus* means *clean.*

pyramid This word comes from the stem of the Greek noun πυραμίς, πυραμίδος, *a pyramid.* It was also used by mistake in the Middle Ages to mean *cone;* so, for example, Albertus Magnus (1193–1280), in his commentary on Euclid's *Elements,* speaks of the ellipse as *sector pyramidis.* The pyramids of Egypt are square-based, or four-sided, but one can generalize the notion so that a pyramid may have any polygon for its plane sections. One produces an n-sided pyramid by taking any regular n-gon inscribed in a circle of radius r and picking a point P at an altitude a above the center of the circle. The pyramid is produced by drawing all the lines connecting P to the points of the given n-gon. If $n = 8$, the octagon-based pyramid thereby produced is the model for the baptistery of San Giovanni in

Florence. The volume of the pyramid is $[nr^2 sin\ (2\pi/n)]a$, and its centroid is $a/4$ up the axis, the same as in the case of a cone.

Pythagorean This is the English adjective formed from the name of the Greek mathematician Πυθαγόρας on the analogy of Latin adjectives ending in *-eanus, -a, -um*. The correct Latin adjective is *Pythagoreus, -a, -um*. A *Pythagorean triplet* is a triplet *a*, *b*, *c* of positive integers such that $a^2 + b^2 = c^2$. The triplet is *primitive* if *a*, *b*, and *c* have no common factor. The primitive Pythagorean triplets were studied by the Babylonians, as we know from the Plimpton cuneiform tablet number 322 in the collection of Columbia University. They are given by $c^2 = p^2 + q^2$, $b^2 = p^2 - q^2$, and $a = 2pq$, where *p* and *q* are relatively prime positive integers such that $p > q$ and exactly one of *p* and *q* is odd.

Q

Q.E.D. This is the abbreviation for the Latin *Quod erat demonstrandum*, which was a Latin translation of the way Euclid ended the proofs of his theorems, ὃ ἔδει δεῖξαι, "This is what it was necessary to prove." Constructions ended with ὃ ἔδει ποιῆσαι, "This is what it was necessary to do," which became *Quod erat faciendum* or Q.E.F. The earliest translators, however, did not use Q.E.D. or Q.E.F. Instead, we find such phrases as *Quod oportebat facere* (Boëthius), *Et hoc est quod proposuimus* (Adelard of Bath), *Et hoc est quod demonstrare voluimus* (Gerard of Cremona), and *Quod proposuimus* (Johannes de Tinamue).

Q.E.F. See the previous entry.

quadri- or quadr- The Latin noun *quadra, -ae* (also *quadrum, -i*) means *a square*. It is related to the adjective *quattuor*, which means *four*. This prefix is therefore correctly attached to other words of Latin origin to convey the idea of *four*.

quadrangle This word is derived from the Latin *quadrangulus*, a plane figure having four angles. See the entries **quadri-** and **angle**.

quadrant This is the stem of the Latin noun *quadrans, quadrantis*, which means *a fourth part*.

quadratic The Latin verb *quadro, quadrare, quadravi, quadratus* means *to make a square* (quadrum). The past participle *quadratus* means *made into a square* or *squared*. The correct English adjective should therefore be *quadrate. Quadratic* is a mistake produced by adding the Greek adjectival suffix *-ic* to the stem of a Latin participle.

quadratrix This word means *squarer, that which squares*, and is the name of a curve used by Hippias (fifth century B.C.) to square the circle, from the Latin *quadro, to square*. The feminine form *quadratrix* is preferred over the masculine form *quadrator* since the name is supposed to be in apposition to *linea curva* (*curved line*) understood. The quadratrix may be defined in the following manner: Consider the square *ABCD* in the first quadrant, whose side *AB* is on the *y*-axis, whose side *AD* is on the *x*-axis, and whose sides each have length *a*. The vertical line segment *BC* descends at uniform speed to the base *AD* at the same time that the side *AB* revolves at uniform angular speed about *A* to the same base *AD*. The locus of intersection of the two moving lines is the quadratrix of Hippias. The polar equation of the quadratrix is $r = 2a\theta / \pi \sin \theta$.

quadrature This is derived from the future active participle *quadraturus, -a, -um* of the verb *quadro, quadrare, to make into a square*.

quadric This word is a mistake, the result of adding a Greek suffix to the root of a Latin noun. The mistake comes from the analogy of *cubic*, but *cube* is of Greek origin.

quadrifolium This well-formed word is the composition of the Latin prefix *quadri-* (from *quattuor, four*) and the noun *folium, leaf*.

quadrilateral This word is composed of the Latin prefix *quadri-* (from *quattuor, four*) and the adjective *lateralis*, which means *pertaining to a side (latus)*.

quadrivium This word is derived from the numeral *quattuor, four*, and the noun *via, road*. It is the Latin name for an intersection of four roads. The quadrivium consisted of the four mathematical subjects of the seven liberal arts, geometry, arithmetic (number theory), astronomy, and music.

quadruple The Latin adjectives *quadruplex, quadruplicis* and *quadruplus* both mean *fourfold*. The second is the parent of *quadruple*. The neuter *quadruplum* as a noun means *four times the amount*.

quantifier This word is a modern invention. The English suffix *-fy* is the remnant of the Latin verb *facere*, which means *to make*. The English suffix *-fier* superimposes on this the additional suffix of agent *-er* so as to convey the meaning *the one who makes*. The word *quantifier* is therefore supposed to mean *the one who makes something as much as it is*.

quantity The Latin interrogative adjective *quantus* means *how much?* From it was derived the noun *quantitas, quantitatis* with the meaning *amount*. This became the French *quantité*, which gave birth to the English *quantity*.

quarter This word is the offspring of the Latin adjective *quartarius*, which means *the fourth part of* something. The intermediate word was the French *quartier*.

quartic The practice of adding the adjectival suffix *-ic* of Greek origin to the stem of Latin ordinal numbers is a mistake that has produced the low words *quartic, quintic, sextic*.

quartile This modern technical term is a noun, although the suffix *-ile* is a remnant of the Latin adjectival suffix *-ilis, -ile*. The Latin adjective *quartus* means *fourth*.

quasi The Latin adverb *quasi* means *almost* or *as if.*

quasianalytic The Latin adverb *quasi* means *almost* or *as if. Analytic,* however, is of Greek origin. The combination is therefore low. It is intended to mean *almost analytic.*

> For a sequence of positive numbers M_1, M_2,... and a closed interval $[a,b] = I$, the *class of quasi-analytic functions* is the set of all functions which possess derivatives of all orders on I and which are such that for each function f there is a constant k such that
>
> $$|f^{(n)}(x)| < k^n M_n \text{ for } n \geq 1 \text{ and } x \in I,$$
>
> provided this set of functions has the property that if f is a member of the set and $f^{(n)}(x_0) = 0$ for $n \geq 0$ and $x_0 \in I$, then $f(x) \equiv 0$ on I. (James and James, *Mathematics Dictionary*, Students edition, D. Van Nostrand Company, Inc., Princeton, New Jersey, 1964, p. 12)

quasiperfect The Latin adverb *quasi* means *almost* or *as if.* See the entry **perfect**. The positive integer n is a *quasiperfect number* if its proper divisors add up to $n + 1$. This is an example of a definition of something that is not known to exist.

quaternion The Latin distributive numeral *quaterni, -ae, -a* means *four at a time.* St. Jerome used the noun *quaternio, quaternionis* to translate the Greek τετράδιον, *a group of four*, *a guard of four soldiers*, in Acts XII 4:

> *Quem cum apprehendisset, misit in carcerem, tradens quattuor quaternionibus militum custodiendum, volens post Pascha producere eum populo.*

> And when he had apprehended him, he put him in prison, and delivered him to four quaternions of soldiers to keep him; intending after Easter to bring him forth to the people.

That there were four quaternions probably means that each quaternion had a six-hour watch. Two of them were chained to the prisoner and the remaining two watched outside the cell. When Wyclife translated Acts XII 4 into English, he used *quaternions* for the Latin *quaternionibus*, an example of a Latinism common in early translators of the Vulgate. Hamilton used the word for his new kind of number in 1843. A quaternion is

a symbol of type $x = x_0 + x_1i + x_2j + x_3k$, where x_0 and the coefficients of i, j, k are real numbers. Scalar multiplication is defined by $cx = cx_0 + cx_1i + cx_2j + cx_3k$; the sum of x and $y = y_0 + y_1i + y_2j + y_3k$ is

$$x + y = (x_0 + y_0) + (x_1 + y_1)i + (x_2 + y_2)j + (x_3 + y_3)k;$$

the product xy is computed by formally multiplying x and y by use of the distributive law and the conventions

$$i^2 = j^2 = k^2 = -1$$

$$ij = -ji = k, jk = -kj = i, ki = -ik = j.$$

The quaternions satisfy all the axioms for a field except the commutative law of multiplication. (James and James, *Mathematics Dictionary*, Students edition, D. Van Nostrand Company, Inc., Princeton, New Jersey, 1964, p. 319)

queue This French word is the corruption of the Latin *cauda*, which means *tail*. Queuing theory studies the distribution of waiting times to obtain a service. The subject was developed by Erlang (1878–1929), although its basis lies in the fact that the random variable whose value is the waiting time until the occurrence of a rare event has the exponential distribution of probabilities, where the number of occurrences of the rare event in a unit time is assumed to have Poisson distribution.

quintic The practice of adding the adjectival suffix *-ic* of Greek origin to the stem of Latin ordinal numbers is a mistake that has produced the low words *quartic, quintic, sextic* pertaining to the polynomial equations of degree 4, 5, and 6, respectively.

quintuple The Latin verb *quinquiplico, quinquiplicare* is used by Tacitus in Book II of his *Annals*, section 36, to mean *to make fivefold*. The two parts are *quinque, five*, and *plico, plicare, plicavi, plicatus, to fold*. Much later the associated adjective *quintuplex, quintuplicis* is found meaning *fivefold, five times as many*, as well as a later form of the aforementioned verb *quintuplico, quintuplicare*. The English noun is derived from the late Latin adjective.

quotient The Latin interrogative adverb *quotiens* means *how often?* The *s* was changed to *t* by people used to Latin third- and fourth-conjugation participles ending in *-iens, -ientis*.

R

radian This modern word was formed as if the Latin adjectival suffix *-anus* had been added to the stem of the Latin noun *radius* to produce the adjective *radianus* with the meaning *pertaining to the spoke of a wheel*, but there is no such Latin word. The radian measure of an angle is due to the scientific system of notation introduced at the time of the French Revolution. An angle measures θ radians if the length of the arc it subtends is θ radii.

radical The Latin noun *radix, radicis* means *root*. The addition of the adjectival suffix *-alis* to the stem produced the late adjective *radicalis* with the meaning *pertaining to the root, having roots*.

radicand This appears to be the stem of a gerund from a Latin verb *radico, radicare*. This verb is intransitive and means *to set down roots, to take root*. Sometime in the sixteenth century, it took on the meaning *to take the root of a number*. *Radicandus, -a, -um* would then mean *[a number] whose root must be taken*.

radius This is a Latin noun with the meaning *staff, rod, stake, beam of light, the spoke of a wheel*. It was not used in the early Latin translation of Euclid for what we now call the *radius* of a circle; they translated the Greek διάστημα literally by *distantia*.

radius vector The Latin plural is *radii vectores*; in English we say radius vectors. This is another name for the *position vector R(t)* that defines a trajectory as a function of the parameter time.

radix The Latin noun *radix, radicis* means *root*. Radix is the translation of the Arabic word جذ, *root*, used by al-Khowarizmi for the solution of an equation.

rate The Latin verb *reor, reri, ratus* means *to reckon, calculate*. The English word is derived from the third principal part. The phrase *pro rata [sc. parte]* means *proportionally*.

ratio This is the Latin word for *reason*, used by the ancients to translate the Greek noun λόγος when it meant *reason* (as opposed to *word*), *proportion*. The noun *ratio* is derived from the third principal part of the verb *reor, reri, ratus*, which means *to think*. If a line is divided at a point into two parts in such a way that the ratio of the larger part to the whole line is the same as the ratio of the smaller part to the larger part, then the common ratio is called the *golden ratio* because the rectangle whose base and altitude are the larger and smaller parts created by such a division is, according to those competent to have an opinion on such a matter, the most aesthetically pleasing of all rectangles that may be formed by dividing the given line and taking the parts to be the dimensions. The golden ratio is $(-1 + 5^{1/2})/2$, which is approximately .618034. The Greek letter φ has traditionally been used for it.

rational The addition of the adjectival suffix *-alis* to the root of the noun *ratio* produced the adjective *rationalis* with the meaning *pertaining to the reason, regarding a proportion*. A rational number is the quotient of two integers; a rational function is the quotient of two polynomials.

rationalize After the adjective *rationalis* had acquired its technical meaning in mathematics, the verb *rationalizo, rationalizare* was coined from it in a natural manner with the meaning *to make rational*. See the entry on **-ize** above.

ray This is the abbreviation of the Latin noun *radius*, which means *a staff, rod, stake, beam of light*.

reaction This word was invented in France in the early seventeenth century from the Latin prefix *re-* and the Latin noun *actio*. There never was a noun *reactio, reactionis* in use in Latin.

real The Latin noun *res, rei* means *thing*. The addition of the adjectival suffix *-alis* to its stem produced the adjective *realis* with the meaning *pertaining to something*. In the law, *real property* is land, called *real* because unlike paper money, it cannot be wiped away by the collapse of the government.

R.E.A.L. This is an acronym for *Research Experiences for All Learners*, as in the title of the book *Keeping It R.E.A.L.*, published by the Mathematical Association of America. It is an example of the habit of giving silly titles to mathematics books, a sign of the deterioration of English usage among those professionally responsible to know better. Examples of such titles, some merely comical, others truly grotesque, include *Expeditions in Mathematics*, *Rediscovering Mathematics*, *Calculus Deconstructed*, *Invitation to Complex Analysis*, *Excursions in Classical Analysis*, *Mathematical Time Capsules*, *Calculus: An Active Approach with Projects*, *Aha! Solutions*, *Who Gave You the Epsilon?*, *Uncommon Mathematical Excursions*, *Biscuits of Number Theory*, *Proofs That Really Count*, *A Garden of Integrals*, *Topology Now!* These titles were all found in the *2011 Fall and Winter Catalogue of MAA Books*. The *2012 Annual Catalogue of Mathematical Association of America Books* also has many examples of books whose titles give no helpful information about what the book is about. What is one to make of *Icons of Mathematics* or *Beautiful Mathematics*? How about *A Mathematician Comes of Age*, *Expeditions in Mathematics*, *Mathematics Galore!*, *She Does Math!*, *N is a Number*, *Hands on History*, *Math through the Ages*, and *When Less is More?* Notice the current fad of putting exclamation points and question marks in titles. One title, *The Great π/e Debate*, uses the slash symbol in an undefined and sloppy manner not expected or to be allowed among mathematicians; the authors are not talking about π divided by e. Several titles use the cataloguese *math* instead of *mathematics*. A related absurdity is to use the word *knot* in place of *not* in titles of books or papers involving knot theory. This is an example of the grossest kind of intellectual limitation.

The concoction R.E.A.L. is also an example of the uncontrolled license with which acronyms are multiplied. I recently received an email identified as coming from IOSSBR. Perhaps I should have known that this was the International Organization for Social Sciences and Behavorial Research. For more on acronyms, see the entry **ANOVA**.

reason This word is derived from the noun *ratio, rationis*, which itself proceeds from the third principal part of the Latin verb *reor, reri, ratus, to think*. *Reason* is the corruption of the root *ration-*. The syllable *ti-* was pronounced *tsi-* in late Latin, and the *t* sound was eventually dropped.

reciprocal The Latin adjective *reciprocus* is formed of the prefixes *re-* (*back*) and *pro* (*forward*) and means *going back and forth*. The addition of the adjectival suffix *-alis* to the stem of what was already an adjective was a mistake on the part of economists of the eighteenth century.

rectangle The adjective *rectangulus* is the Latin equivalent of the Greek ὀρθογώνιος, *right-angled*.

rectify The Latin verb *rego, regere, regi, rectus* means *to guide, direct, rule*. The past participle *rectus*, which meant *guided, directed, ruled* by a natural development then came to mean *straight*, for those under government are assumed to exhibit orderly behavior. The phrase *linea recta* was used by the translators of Euclid to mean *straight line*. The English verb *rectify* is derived from the late Latin creation *rectifico, rectificare, to make straight*. The suffix *-ify* is clumsily used in the creation of English verbs from Latin verbs ending in *-ifico*.

rectilinear The Latin adjective *rectilinearis*, which means *straight-lined*, is compounded of the prefix *recti-* and the adjective *linearis*.

recurrent The Latin verb *recurro, recurrere, recurri, recursus* means *to run* (*curro*) *back* (*re-*), *to hasten back, to revert, return*. Its present participle is *recurrens, recurrentis*, whose stem is the English adjective *recurrent*. The Poincaré recurrence theorem (1890) says that if T is a measure preserving transformation on *[0,1]* and *A* is a measurable set, then for almost all *x* there is a positive integer $n = n(x,A)$ such that $T^n(x) \in A$.

recurring The Latin verb *recurro, recurrere, recurri, recursus* means *to run back, to hasten back, to revert, return*. Its first principal part is the root of the English verb *to recur*.

recursive This word has no existence outside of mathematics. It is formed as if the Latin adjectival suffix *-ivus* was added to the stem of the participle *recursus* to produce the adjective *recursivus*, pertaining to a return.

reduce The Latin verb *reduco, reducere, reduxi, reductus* means *to lead* (*duco*) *back* (*re-*).

reducible This English word was formed by adding the English modification of the Latin suffix *-abilis* to the stem of the verb *reduco*. See the previous entry.

reductio ad absurdum This is the Latin name for a proof by contradiction, literally, *a reduction to absurdity*. It is a paraphrase of the Greek term ἡ εἰς τὸ ἀδύνατον ἀπαγωγή of Aristotle (*Prior Analytics*, I. 7, 29 b 5), which literally means *to lead back to the impossible*.

redundant The Latin noun *unda* means *water, water in motion, a stream, a wave*. The derivative verb *redundo, redundare, redundavi, redundatus* means *to run back or over* (of water), *to stream over, to roll over, to overflow, to be in abundance, to be left over*. Its present participle is *redundans, redundantis*, whose stem is the English adjective *redundant*.

reflection The Latin verb *flecto, flectere, flexi, flexus* means *to bend*, and the verb *reflecto, reflectere, reflexi, reflexus* means *to bend back*. The associated Latin noun *reflexio, reflexionis* is taken from the fourth principal part, so the correct English noun would have been *reflexion*, not *reflection*. The mistake is due to the fact that the verb *reflect* is taken from the first and second principal parts.

reflector This is cant for a student in a mathematics class whose job it is to reflect on what is going on. It belongs to the language of "mathematics education."

reflexive The addition of the adjectival suffix *-ivus* to the stem of the participle *reflexus* produces the adjective *reflexivus* with the meaning *pertaining to a bending back*.

refraction The Latin verb *frango, frangere, fregi, fractus* means *to break*. The associated verb *refringo, refringere, refregi, refractus* means *to break back, to break up, to break open*. From the stem of the fourth principal part is produced the noun *refractio, refractionis, a breaking back*.

region The Latin verb *rego, regere, rexi, rectus* means *to guide or direct*. The associated noun *regio, regionis* is *a direction, a line, a boundary line, a territory marked out with lines*. A non-empty subset of R^n is a *region* if it is open and connected.

regress The use of this word as a verb "to regress X on Y" is common in statistics books, but is awkward and even ugly. In correct prose, *regress* is an intransitive verb.

regression The Latin verb *regredior, regredi, regressus* means *to walk back*. The noun *regressio, regressionis* is formed from the last principal part and means *a walking back*.

> The latter term "regression" appears in his [*sc.* Galton's] Presidential address made before Section H of the British Association at Aberdeen, 1885, printed in Nature, September, 1885, pp. 507–510, and also in a paper "Regression towards mediocrity in hereditary stature," Journal of the Anthropological

Institute, 15, 1885, pp. 264–263. (N. R. Draper and H. Smith, *Applied Linear Regression Analysis*, second edition, John Wiley & Sons, New York, 1981, p. 4)

regula [loci] falsi This is a technical term meaning *the rule of false [position]*. The noun *regula* is derived from *rego, to direct*, whereas *falsi* is the masculine genitive singular of the past participle *falsus* of the verb *fallo, to deceive*.

regular The Latin noun *regula* means *a straight length, a ruler*. It is derived from the verb *rego, regere, rexi, rectus*, which means *to direct*.

regulus The Latin noun *rex, regis* means *king*; it is derived from the verb *rego, regere, rexi, rectus, to guide, to rule over*. The addition of the diminutive suffix *-ulus* to the stem of the noun produced the word *regulus, a little king*.

related The Latin verb *refero, referre, retuli, relatus* means *to carry back*. From its first principal part we get our verb *to refer*, and from its fourth principal part we derive our verb *to relate*.

relation This word is derived from the stem of the Latin noun *relatio, relationis*, which is derived from the fourth principal part of the verb *refero*. See the previous entry.

relative The Latin adjectival suffix *-ivus* has been added to the stem of the past participle *relatus* of the verb *refero* to produce the adjective *relativus* with the meaning *having reference or relation*.

remainder The Latin verb *remaneo, remanere, remansi, remansus* means *to stay behind, to stay where one is while others move*. It is the compound of the prefix *re-* (*back, again*) and the verb *maneo* (*to stay*). The *d* was added by the French before the word came into English.

removable, removeable This word is produced by adding the Latin adjectival suffix *-abilis* to the stem of the first principal part of the verb *removeo*. The spelling with the *e* is preferable since the Latin verb is second conjugation. The spelling without the *e* was the result of the activity of people who were afraid that the word might be mispronounced as *removeeble*.

rencontre This is the French verb *to match, to meet*. It is compounded of the Latin prefix *re-* and the Latin prepositions *in* and *contra*. From the two prepositions there proceeded the late Latin and Italian verb *incontrare, to meet*. The word *rencontre* gives its name to a famous problem in the theory of probability. If one takes two identical decks of *n* cards numbered *1* to *n*, shuffles them separately, and then arranges them at random in two adjacent columns, what is the probability that there will be exactly *r* matches, $r \leq n$? The required probability is

$$[1/2! \; - \; 1/3! \; + \; 1/4! \; - \; 1/5! \; + - \; \cdots + (-1)^{n-r}/(n-r)!]/r!$$

The probability of at least one match is

$$1 - 1/2! \; + \; 1/3! \; - \; 1/4! \; + \; 1/5! \; - + \; \cdots \; + (-1)^{n-1}/n!,$$

which is the partial sum of the alternating harmonic series, which rapidly converges to *(e − 1)/e*; whether the decks have seven or seven trillion cards, the probabilities are the same to four decimal places.

replication The Latin verb *replico, replicare, replicavi, replicatus* means *to fold (plico) back (re-), unroll, review*. From its fourth principal part is derived the noun *replicatio, replicationis* with the meaning *a folding back*.

representation The Latin adjective *praesens, praesentis* means *to be in front or before*. From it is derived the verb *repraesento, repraesentare, repraesentavi, repraesentatus* with the meaning *to make present again*. From the stem of its fourth principal part comes the noun *repraesentatio, repraesentationis*, a vivid or lively presentation.

repunit This is a ludicrous word, the comical abbreviation of *repeated unit*. According to Schwartzman, the word is due to Albert Beiler.

One should not do this sort of thing unless one's purpose is to be ridiculous.

residual The Latin *resido, residere, resedi* means *to sit (sedeo) back (re-), remain seated*. From it is derived the adjective *residuus, -a, -um* meaning *left behind, remain over*. The superfluous addition of the adjectival suffix *-alis* produced the comical superadjective *residualis*, whose stem became the English word.

> Almost all the greatest discoveries in astronomy have resulted from the consideration of what we have elsewhere termed RESIDUAL PHENOMENA, of a quantitative or numerical kind, that is to say, of such portions of the numerical or quantitative results of observation as remain outstanding and unaccounted for after subducting and allowing for all that would result from the strict application of known principles. (Sir John F. W. Herschel, Bart. K. H. in *Outlines of Astronomy*, Lea and Blanchard, Philadelphia, 1849, p. 548, quoted in N. R. Draper and H. Smith, *Applied Linear Regression Analysis*, second edition, John Wiley & Sons, New York, 1981, p. 141)

residue This word comes from the French adjective *residu*, itself derived from the Latin adjective *residuus, -a, -um*. The plural *residues* was in Latin the neuter plural *residua*. See the previous entry.

resolution This word is derived from the Latin noun *resolutio, resolutionis, a loosening*, which is formed from the fourth principal part of the verb *resolvo*. See the following entry.

resolvent The Latin verb *solvo, solvere, solvi, solutus* means *to set free*. The addition of the prefix *re-*, which has the force of *again, back*, produces the compound verb *resolvo, resolvere, resolvi, resolutus* with the meaning *to set free and so put back in its original state, unbind, loosen*. Its present participle is *resolvens, resolventis*, whose stem is the English word *resolvent*.

resonance The Latin verb *sono, sonare, sonavi, sonatus* means *to sound*, and the compound verb *resono, resonare, resonavi, resonatus* means *to sound (sono) back (re-)*. The noun *resonance* is the transfiguration of the

Latin *resonantia, -ae, an echo*. It is the property of certain systems to vibrate with greater amplitude at certain frequencies.

> One of the questions of greatest interest is the study of nonlinear [*sc.* differential] equations which are almost linear near one or more eigenvalues. These cases are known under the name of *resonance problems*. (Piccinini, Stampacchia, and Vidossich, *Ordinary Differential Equations in R^n: Problems and Methods*, Springer-Verlag, New York, 1984, p. 258)

respectively The Latin verb *respicio, respicere, respexi, respectus* means *to look back*. The addition of the adjectival suffix *-ivus* to the stem of the fourth principal part produces the late adjective *respectivus*, from which the adverb *respective* was formed whence proceeded the English adjective *respective*.

restriction The Latin verb *stringo, stringere, strinxi, strictus* means *to draw tightly together, to bind, tie*. The addition of the prefix *re-* produces the compound verb *restringo, restringere, restrinxi, restrictus* with the meaning *to bind back, draw back, confine, restrict*. The English word *restriction* is the stem of the associated noun *restrictio, restrictionis* with the same meaning.

result The Latin verb *salio, salire, salui, saltus* means *to jump*. The addition of the prefix *re-* results in the compound verb *resilio, resilire, resilivi, resultus, to jump back*. From this word is formed the frequentative verb *resulto, resulatare, resultavi, resultatus* meaning *to spring back constantly, to rebound constantly*, from which is derived the English verb and noun *result*.

resultant This is the stem of the present participle *resultans, resultantis* of the verb *resulto*. See the previous entry.

retract The Latin verb *traho, trahere, traxi, tractus* means *to drag*. The addition of the prefix *re-* produces the compound verb *retraho, retrahere, retraxi, retractus* meaning *to draw (traho) back (re-)*. From its fourth principal part is derived the late Latin noun *retractio, retractionis, a drawing back*.

retraction This noun is the stem of the late Latin noun *retractio, retractionis*. See the previous entry.

reverse The Latin verb *verto, vertere, verti, versus* means *to turn*. The addition of the prefix *re-* adds the force of *back* and produces the compound verb *reverto, revertere, reverti, reversus* meaning *to turn* (*verto*) *back* (*re-*). The English verb comes from the fourth principal part.

reversion The Latin noun *reversio, reversionis* means *a turning back before the end of a journey*. See the previous entry.

revolution The late Latin noun *revolutio, revolutionis* is derived from the fourth principal part of the verb *revolvo*. See the following entry.

revolve The Latin verb *volvo, volvere, volvi, volutus* means *to roll*. The addition of the prefix *re-* imparts the notion of *back* and so produces the compound verb *revolvo, revolvere, revolvi, revolutus* meaning *to roll* (*volvo*) *backwards* (*re-*).

rho This is the letter of the Greek alphabet corresponding to our R, r.

rhodonea The Greek noun ῥόδον means *a rose*. From it were formed the nouns ῥοδών, ῥοδῶνος, *a bed of roses*, and ῥοδωνιά, *a garden of roses*, and also simply *a rose*. Grandi (1671–1742), for whom the Latin word *rosa* was not good enough, took the Greek word for the name of his curve. See the entry **rose**.

rhombohedron See the entries **rhombus** and **-hedron**. *Rhombohedron* is a good example of the proper way of making new words on the analogy of classical examples.

rhomboid *Rhombus* is the Latin transliteration of the Greek noun ῥόμβος or ῥύμβος, which means *anything that can be twirled or that is unsteady*, from the verb ῥέμβω, *to twirl, to be unsteady, to act at random*. The addition of the suffix -ειδής, -ειδές (from the noun εἶδος, *that* *which is seen, the form, shape, figure*) to the stem with the connecting vowel *o* produces the Greek adjective ρομβοειδής, ρομβοειδές, which means *like a rhombus.*

rhombus This is the Latin transliteration of the Greek noun ρόμβος, which means *anything that can be twirled*, from the verb ρέμβω, *to twirl, to be unsteady, to act at random.*

rhumb This is the corruption of the Latin noun *rhombus* or *rhombus*, the transliteration of the Greek ρόμβος or ρύμβος.

right This is cognate with the Latin adjective *rectus, right*, from *rego, to direct.*

rigid The Latin verb *rigeo, rigere* means *to be stiff, to stiffen*. It is connected with the verb *frigeo, frigere*, which means *to be cold*. Both are cognate with the Greek verb ριγέω, *to shiver from the cold*. From *rigeo* proceeded the adjective *rigidus*, which means *stiff*. The English *rigid* is the stem of the Latin adjective.

robust This word has become mathematical cant from overuse. *Robur* in Latin means *oak tree*, and the associated adjective *robustus* means *strong like an oak tree*. However, the word *robust* is now used frequently as a catch-all word whenever some positive attribute is to be attributed to something. For example, the president of a college recently assured the faculty that he was implementing a new "robust structure" in the bureaucracy; all this means is that, in his opinion, he has improved the organization of the administration. Furthermore, when Hurricane Irene approached the Atlantic coast, an authority who was being interviewed on the Weather Channel announced that a "robust plan" was in place to prevent the cell phone system from crashing during the storm.

Roman numerals This was the method of writing the natural numbers among the Romans. The Roman numerals were not used as adjectives for the ordinal numbers, so it is incorrect to write "the XX Olympiad." Their use nowadays adds a sense of dignity to inscriptions, and they may be used for volume numbers of periodicals or series and chapter numbers of books. Otherwise their use is an affectation.

rose The Latin noun *rosa* means *rose*. The *roses of Grandi* were introduced by the Camaldolese Benedictine mathematician Luigi Guido Grandi in 1728 in his book *Flores Geometrici*, where they are called *rhodoneae*; they are the polar curves with equation $r = a \cos n\theta$, where n is a positive integer. If n is odd, the rose has n leaves. If n is even, the rose has $2n$ leaves. Thus, there are no roses with 2, 6, 10, 14, 18,… leaves. The area of the region enclosed by the petals of the rose is $\pi a^2/4$ if n is odd, and $\pi a^2/2$ if n is even. The rose may be produced in the following fashion: Let n be a fixed positive integer. For every angle θ, the ray that makes an angle $n\theta/2$ with the initial line intersects the circle C with equation $r = a \cos \theta$ at some point C. The perpendicular from C intersects the initial line at D. Let B be the point $(a,0)$. There is a point P on OC (or its extension) such that $OP = OD - DB$. The rose of Grandi is the locus of the point P.

rosette The addition of the French suffix *-ette* to the stem of *rose* produced the diminutive *rosette, a little rose*. The Romans would have said *rosina*.

rotate The Latin noun *rota* means *wheel*. From it was formed the verb *roto, rotare, rotavi, rotatus* with the meaning *to whirl around, to cause to spin*. Our verb is derived from the fourth principal part.

rotation This word is the stem of the noun *rotatio, rotationis*, which is derived from the fourth principal part of the verb *roto*. See the previous entry. If the *x*- and *y*-axes of the Cartesian plane are rotated through an angle θ, then the previous coordinates *(x,y)* of a point are changed to *(x',y')* where

$$x = x' \cos \theta - y' \sin \theta$$

$$y = x' \sin \theta + y' \cos \theta.$$

rotund This word is the stem of the Latin adjective *rotundus*, which means *round, circular*, from *rota, wheel*.

roulette The French word *roue* is the metamorphosis of the Latin *rota*, which means *wheel*. The addition of the suffix *-elle* produced the diminutive *rouelle, a little wheel*. Someone who forgot that this noun was already a diminutive superimposed the additional suffix *-ette* to produce the superdiminutive *roulette, a little little wheel*.

round This word is the transformation of the Latin adjective *rotundus*. The dropping of the letter *t* was inherited from the French *rond*.

Rousseau's problem In the *Confession of Faith of the Savoyard Vicar*, one finds this passage:

> And yet if any one were to tell me that a number of printer's types, jumbled promiscuously together, had arranged themselves in the order of the letters composing the *Aeneid*, I certainly should not deign to take one step to verify or disprove such a story. (The Harvard Classics, *French and English Philosophers*, Easton Press Edition, 1994, pp. 259–260)

Rousseau's problem is to calculate the probability that the *Aeneid* would be produced if a letter were selected from the alphabet at random and with replacement a number of times equal to the total number of letters in the poem. According to the *law of large numbers*, however, if this process were continued indefinitely, the poem must eventually be produced with probability one.

ruin This word is the stem of the Latin noun *ruina*, which means *the state of financial collapse*. The instructive game of gambler's ruin is part of the curriculum of the theory of probability. Peter has *a* dollars and Paul has *b* dollars. A referee takes up a fair coin and proceeds to toss it. Whenever it comes up heads, Peter takes \$1 from Paul. Whenever it comes up tails, Paul takes \$1 from Peter. The game is over when one of the two is ruined. (By the weak law of large numbers, the probability is *1* that someone will be ruined.) The probability that Peter will ruin Paul is $a/(a + b)$, and the probability that Paul will ruin

Peter is $b/(a + b)$. The expected number of tosses until someone is ruined is ab. This last result was considered a paradox by many, for if $a = 1$ and $b = 1,000,000$, one expects Peter to be ruined right away. The surprise is to be explained in the same manner as the Petersburg paradox, *q.v.*

rule This is the corruption of the Latin noun *regula*. See the entry **regula** above.

ruler The suffix of agency *-er* has been added to the verb *rule* to produce *ruler*, that which rules.

S

saddle The Latin word *sella* means *saddle*. It is related to the noun *sedile*, which means *seat*. Both *sella* and *sedile* are derived from the verb *sedeo, sedere, sedi, sessus*, which means *to sit*.

salient The Latin verb *salio, salire, salui, saltus* means *to jump*. Its present participle is *saliens, salientis*, whose stem is the English adjective *salient*.

saltus This is the Latin fourth-declension noun derived from the fourth principal part of the verb *salio*. See the entry **salient** above.

sample This word is the corruption of the Latin noun *exemplum*. See the entry **example**.

sampling This word is the English gerund of *sample*. See the preceding entry.

satisfy The Latin verb *satisfacio, satisfacere, satisfeci, satisfactus* means *to do (facio) enough (satis) for, to make amends to*. The *c* was already lost by the French, who used *faire* for *facere*.

saturated The Latin adverb *satis* means *enough*, and the related adjective *satur, satŭra, satŭrum* means *full of food, sated*. From this adjective proceeded the verb *saturo, saturare, saturavi, saturatus* with the meaning *to glut, to fill*, whence came our verb *to saturate*.

> If $\{X, \mathcal{B}, \mu\}$ is a measure space, we say that a subset E of X is *locally measurable* if $E \cap B \in \mathcal{B}$ for each $B \in \mathcal{B}$ with $\mu B < \infty$.... The measure μ is called *saturated* if every locally measurable set is measurable. (Royden, p. 221)

scalar The Latin noun *scala* means *a staircase*. It is related to the verb *scando, scandere, scandi, scansus*, which means *to climb*. The addition of the adjectival suffix *-alis* to the stem of the noun produced the word *scalaris*, with the meaning *pertaining to a staircase*.

scale *Scala* is the Latin word for *staircase*.

scalene This is the metamorphosis of the Greek adjective σκαληνός, which means *limping*, formed from the verb σκάζω, *to limp*. It went over into late Latin as *scalenus*.

scattergram The Greek verb σκεδάννυμι means *to scatter* and is probably the ultimate source of the English verb. The noun *gram* is derived from the Greek nouns τὸ γράμμα, *a letter*, and ἡ γραμμή, *a stroke in writing, a line*.

science The Latin verb *scio, scire, scivi, scitus* means *to know*, and the derived noun *scientia* means *knowledge, learning*.

scientific The late Latin adjective *scientificus, -a, -um* was derived from the noun *scientia*, which means *knowledge*, and the verb *facio*, which means *to make, to do*; the *-ic* is not from -ικός.

secant The Latin verb *seco, secare, sectus* means *to cut*. Its present participle *secans, secantis* means *cutting*, and its root is the English word *secant*. The secant of the angle θ is the length of the line segment from the origin to the point *(1, tan θ)*. The line segment in question is called the *secant line* because it *cuts* through the unit circle.

sech This is the standard abbreviation for the hyperbolic secant function: *sech x = 1/cosh x*. The abbreviation stands for *cosecans hyperbolica*. Someone somewhere is probably pronouncing it *sĕch*, but it should be read *hyperbolic secant*.

second The Latin verb *sequor, sequi, secutus* means *to follow*. From this verb proceeded the adjective *secundus* with the meaning *following after the first*. The use of the noun *second* as the division of time is derived from the Latin phrase *secunda minuta*, the *second minute*.

secondary The Latin adjective *secundarius* means *second-rate*. For example, Suetonius speaks of the second-rate bread (*panis secundarius*) that the emperor Tiberius provided for Rome.

section The Latin verb *seco, secare, secui, sectus* means *to cut*. From the fourth principal part comes the noun *sectio, sectionis* with the meaning *a cutting, a division into parts*. The English word is the stem of this noun.

sector The Latin verb *seco, secare, secui, sectus* means *to cut*. From the fourth principal part comes the noun of agent *sector*, which means *he who cuts*. The use of this noun to mean a piece of a circle bounded by two radii and the subtended arc is a mistake. Latin nouns ending in *-or* refer to people who do things; if it is felt necessary to have a special word for the female agent, the suffix *-trix* is used. The sector of a circle is the region bounded by two radii and the circumference. If the central angle of a sector in a circle of radius *r* is *θ* radians, then the area of the sector is $r^2\theta/2$.

secular The Latin noun *saeculum* means *century*. Its etymology is dubious. The related adjective *saecularis, -e* means *pertaining to a century*. If *A* is a matrix whose entries are taken from some field, the *secular equation of A* is the equation *det (A − λI) = 0*. The modern name for *secular equation* is *characteristic equation*. The equation was called *secular* because of an application to the study of planetary motion in astronomy, where small perturbations over the course of a century in a planet's orbit were denominated *secular*.

segment The Latin noun *segmentum* means *a piece cut off*. It is related to the verb *seco, to cut*. A *segmental arc* is a path consisting of the union of line segments and is discussed by Knopp (Part I, p. 15).

self-adjoint transformation Here are combined in one phrase three words, one each of Anglo-Saxon, Latin, and Greek origin. The prefix *self-* is an Anglo-Saxon word. Its use with nouns and adjectives of Latin or Greek origin marks a phrase as modern. For the etymologies, see the entries **adjoint** and **transformation**.

semester This is the name for the most common school term in America. The Latin adjective *semestris, -e* means *lasting six (sex) months (menses)*.

semi- The Latin prefix *semi-* means *half* and is cognate with the Greek prefix ἡμι-. Which prefix is used depends on the word to which it is attached. The use of the prefix to indicate *nearly* or *almost* is in the Latin tradition of phrases like *semisepultus, half-buried*, and *semisomnus, half-asleep*.

semi-algebra Royden (*Real Analysis*, second edition, Macmillan, 1970, p. 259) defines this word thus:

> We say that a collection C of subsets of X is a *semialgebra* of sets if the intersection of any two sets in C is again in C and the complement of any set in C is a finite disjoint union of sets in C.

This is quite a bold use of the prefix; it serves merely to indicate some loosening of the requirements for a collection to be an algebra. See the entries **semi-** and **algebra**.

semicircle This is the metamorphosis of the Latin noun *semicirculus* of the same meaning. The prefix *semi-* here has its strict meaning. See the entries **semi-** and **circle**.

semi-closed Halmos defines a *semiclosed interval* to be an interval of the form *[a,b)*, that is, closed on the left and open on the right. See his *Measure Theory*, Van Nostrand Reinhold Company, 1950, pages 32–33. See the entries **semi-** and **closed**.

semi-conjugate axis See the entries **semi-**, **conjugate**, and **axis**. The conjugate axis of a hyperbola is the line segment through the center of the hyperbola perpendicular to and bisected by the transverse axis of the hyperbola, and whose length is $2a(e^2 - 1)^{1/2}$, where $2a$ is the length of the transverse axis (the distance between the vertices) and e is the eccentricity of the hyperbola; the term *conjugate axis* is also loosely used for the length of this line segment. The semi-conjugate axis is half the length of the conjugate axis.

semi-continuous See the entries **semi-** and **continuous**. The use of *semi-* here found favor in the eyes of Halmos; for the reason why, see the entry **semi-metric**. According to Royden (*Real Analysis*, second edition, Macmillan, 1970, p. 48), an extended real-valued function f is called *lower semi-continuous* at the point y if $f(y) \neq -\infty$ and $f(y)$ is less than or equal to the limit inferior of $f(x)$ as x approaches y. Similarly, f is called *upper-semicontinuous* at y if $f(y) \neq \infty$ and $f(y)$ is greater than or equal to the limit superior of $f(x)$ as x approaches y.

semi-cubical parabola This is the name of the function whose formula is $y = x^{3/2}$. It is neither a parabola nor half a parabola. See the entries **semi-**, **cubical**, and **parabola**.

semi-finite According to Royden (*Real Analysis*, second edition, Macmillan, 1970, p. 220), "A measure μ is said to be *semifinite* if each measurable set of infinite measure contains measurable sets of arbitrarily large finite measure." This is a daring use of the prefix. See the entries **semi-** and **finite**.

semi-group This is an example of a Latin prefix that has become so familiar that it is now attached to all sorts of words to indicate a degree of insufficiency. A semi-group fails to be a group either because it lacks an identity element or because some of its elements lack inverses. The notion of halfness is entirely absent. See the entry **semi-**.

semi-major The *semi-major axis* of an ellipse is half the major axis, that is, half the length of the longest diameter of the ellipse. We have here a true use of the prefix *semi-*. See the entries **semi-**, **major**, and **ellipse**.

semi-metric This word is found in Kelley's *General Topology*, It was condemned by Halmos, who suggested *pseudo-metric*.

> In re semi-metric versus pseudo-metric, I much prefer the latter....Semi-metric...is bad because "semi," meaning "half," hints at the number 2. I would say that "semi" is justified only if there is some hint of duality in sight: e.g., for semicontinuous functions. (Paul Halmos, *I Want to Be a Mathematician*, Mathematical Association of America, 1985, p. 339)

See the entries **semi-** and **metric**.

semi-minor The *semi-minor axis* of an ellipse is half the minor axis, that is, half the length of the shortest diameter of the ellipse. See the entries **semi-**, **minor**, and **ellipse**. We have here a true use of the prefix *semi-*.

semi-norm A *semi-norm* on E^n is a real valued function f that satisfies all the conditions for a norm except that the requirement $f(x_1,\ldots,x_n) = 0 \Rightarrow x_1 = \ldots = x_n = 0$ is replaced by $f(x_1,\ldots,x_n) \geq 0$ for every $(x_1,\ldots,x_n) \in E^n$. This illustrates a misuse of the prefix *semi-*. The function f does not satisfy half the postulates for a norm, nor is it half a norm. Halmos condemned this misuse of the prefix in a letter to Kelley. See the entries **semi-** and **norm**.

semi-perfect A *semi-perfect number* is a natural number that is equal to some of its proper divisors. The prefix here indicates merely a weakening of the condition for a natural number to be perfect, a blameworthy usage. A sign of the precarious standing of the definition is the fact that the term *pseudo-perfect* is also used for the same concept. See the entries **semi-** and **perfect**.

semi-ring In this case the hyphen is indispensible. Otherwise one produces the dreadful looking *semiring*. See the entry **semi-**.

separable The Latin verb *separo, separare, separavi, separatus* means *to disjoin, sever*. The addition of the adjectival suffix *-abilis* to the stem of the first principal part produces the word *separabilis* with the meaning *capable of being severed*. A variables separable differential equation is one of the form $M(x) + y'(x) N(y) = 0$, which can be rewritten in the form $M(x)dx + N(y) dy = 0$ so that the variables x and y are separated.

separation The Latin verb *separo, separare, separavi, separatus* means *to disjoin, sever*. The English noun is the stem of the Latin noun *separatio, separationis, a severance*, formed from the fourth principal part.

separatrix The Latin verb *separo, separare, separavi, separatus* means *to disjoin, sever*. The addition of the suffix of a feminine agent *-trix* to the stem of the fourth principal part produces the noun *separatrix*, which means *she who separates*.

septagon This *vox nullius* is a learned mistake for *heptagon*. People of some education make a certain type of error not committed by the multitude, and this word is an example of one such mistake, *viz.*, the confusion of languages. Knowing that they need a foreign word for *seven*, they take the familiar Latin word instead of the required but unfamiliar Greek word and concoct the hybrid *septagon* on the analogy of the common term *pentagon*. Related absurdities are *automobile, homosexual, neuroscience, sociopath, television*, etc. Any such word may be immediately identified as modern. The word *septagon* appears in Herstein's *Topics in Algebra* (first edition, 1964) on pages 190 and 341, an example of a mathematical Homer nodding. It was corrected to *heptagon* on page 242 of *Abstract Algebra*, Macmillan, New York, 1986.

sequence *Sequor* in Latin means *to follow*, and the associated noun *sequentia* means *continuation*.

sequential See the previous entry. The addition of the adjectival suffix *-alis* to the stem of the noun *sequentia* produces this adjective with the meaning *pertaining to a continuation*.

series The Latin verb *sero, serere, serui, sertus* means *to join together, to put in a row, to connect*. It is related to the Greek εἴρω with the same meaning. (It is not to be confused with the verb *sero, serere, sevi, satus*, which means *to sow, set, plant*.) From it was derived the noun *series*, which means *a row, succession, or chain*.

serpentine This is the name of a plane curve studied by Newton in 1701, which has the shape of a snake, and was therefore called the *linea serpentina* or serpent curve. *Serpentine* by itself is really the transliteration of only the adjective, and it is therefore incorrect to refer to the curve by it alone; it would be better to call the curve *the serpent*. The formula of the serpentine curve is $y = x/(1 + x^2)$.

sesquicentennial *Semis* is Latin for *a half*, and the enclitic *-que* means *and*. From the combination was produced the adverb *sesqui* with the meaning *more by a half*. It was used in compounds like *sesquihora, an hour and a half*, and *sesquimensis, a month and a half*. *Sesquicentennial* is an English invention formed from Latin roots to mean *pertaining to the one hundred and fiftieth anniversary*.

sesquiplicate The Latin adjective *sesquiplex, sesquiplicis* means *taken one and a half times*. It is composed of the adverb *sesqui* meaning *more by a half* and the suffix *-plex* from the verb *plico, plicare, plicavi, plicatus*, which means *to fold*. A *sesquiplicate ratio* is a ratio of the form T^2/d^3, as occurs in Kepler's third law of planetary motion.

set As a noun indicating a collection or group, this word is derived from the third principal part of the Latin verb *sequor, sequi, secutus*, which means *to follow*.

sexagenarian A sexagenarian is a person who has attained the age of sixty. It is derived from the Latin adjective *sexagenarius, viz.,* a sixty-year-old fellow, for the Latin word for sixty is *sexagingta*. It was the age at which a Roman citizen lost the franchise and was disqualified from voting in the assembly. At election time, therefore, the sexagenarians were admonished to keep clear of the bridge over the Tiber, so that the crowds of voters might more easily arrive at the polls. As a result there arose the proverb, *Sexagenarii de ponte!—If you are sixty, get off the bridge!*

sexagesimal The Latin word for sixty is *sexaginta*. The corresponding ordinal number is *sexagesimus* with the meaning *sixtieth*. The superimposition of the adjectival suffix *-alis* on the stem of what is already an adjective produced the word *sexagesimalis* with the meaning *pertaining to the sixtieth part*, which was taken over into English.

sextic The practice of adding the adjectival suffix *-ic* of Greek origin to the stem of Latin ordinal numbers is a mistake that has produced the low words *quartic, quintic, sextic.*

sextuple This English noun was formed on the analogy of *quintuple*. See that entry.

sigma This is the letter of the Greek alphabet corresponding to our S. The capital letter is Σ, and the small case letter is σ. When the σ comes last in a word, it was changed to ς, which is the origin of our letter s. The use of the capital *sigma* to indicate a sum is due to the letter s being the first letter of the Latin word *summa, sum*; our S is merely a corruption of Σ accomplished by careless handwriting. The integral sign $\int$ is another transfiguration of S. The word *sigma* is used as an English mathematical prefix indicating that the idea implied in the following word is being combined with the notion of countable infinity. The following nine entries are examples of this.

sigma-algebra (σ-algebra) This is another name for a *sigma-field*. See that entry below.

sigma-bounded (σ-bounded) See the entry **sigma**. A set that is contained in a sigma-compact subset of a topological space is called *sigma-bounded*.

sigma-compact (σ-compact) See the entries **sigma** and **compact**. A subset of a topological space X is *sigma-compact* if it is the union of a countable collection of compact sets.

sigma-field The noun *field* is of Anglo-Saxon origin. In the theory of probability, the sigma-field is that collection of subsets, called *events*, of the sample space to which one assigns probabilities. A sigma-field of subsets of a set is a non-empty collection that is closed under complementation and the taking of countable unions. See the entry **sigma**.

sigma-finite (σ-finite) See the entries **sigma** and **finite**. A measurable subset of a measure space is of *sigma-finite measure* if it is the union of a countable collection of sets of finite measure.

sigma-homomorphism (σ-homomorphism) See the entries **sigma** and **homomorphism**. Suppose Q is an algebra of subsets of X and B is an algebra of subsets of Y. A function Φ from B to Q is a *(lattice) homomorphism* if $\Phi(Y) = X$, $\Phi(\sim B) = \sim\Phi(B)$ for all $B \in B$, and $\Phi(A \cup B) = \Phi(A) \cup \Phi(B)$ for all A and B in B. If Q and B are sigma-algebras, and Φ has the property that $\Phi(\cup E_i) = \cup \Phi(E_i)$, where the unions are from 1 to ∞, then Φ is called a *sigma-homomorphism*. See Royden, page 318, from which these definitions are taken.

sigma-ideal (σ-ideal) See the entries **sigma** and **ideal**. Suppose $\{X, Q, \mu\}$ is a measure space, and $\mathcal{N}$ is a family of sets in Q with the following properties: i) For all $A \in \mathcal{N}$ and $B \in Q$ such that $B \subseteq A$, we must have $B \in \mathcal{N}$, and ii) If $A_n \in \mathcal{N}$, then $\cup A_n \in \mathcal{N}$. Then $\mathcal{N}$ is called a *sigma-ideal*. See Royden, page 320.

sigma-isomorphism (σ-isomorphism) See the entries **sigma** and **isomorphism**. For the following definition, refer to the entry **sigma-**

homomorphism. A sigma-homomorphism Φ from a Boolean sigma-algebra $\mathfrak{A}$ to a Boolean sigma-algebra $\mathfrak{B}$ is a *sigma-isomorphism* if there is a sigma-homomorphism Ψ from $\mathfrak{B}$ to $\mathfrak{A}$ such that $\Psi \circ \Phi$ is the identity on $\mathfrak{A}$ and $\Phi \circ \Psi$ is the identity on $\mathfrak{B}$. See Royden, page 329.

sigma-ring (σ-ring) See the entry **sigma.** If X is a set, a collection $\mathfrak{A}$ of subsets of X is a *sigma-ring* of subsets of X if $A - B \in \mathfrak{A}$ for all $A, B \in \mathfrak{A}$, and $UE_i \in \mathfrak{A}$ whenever $E_i \in \mathfrak{A}$, $I = 1, 2, 3, \ldots$ See Royden, page 222.

sign This word is the stem of the Latin noun *signum*, which means *a mark or token.*

signature From the Latin noun *signum*, which means *a mark or token*, there was produced the verb *signo, signare, signavi, signatus*, which means *to put a mark upon, inscribe.* The future active participle of this verb is *signaturus, -a, -um,* from whose feminine form *signatura* the noun *signature* was formed.

significance The Latin verb *significo, significare, significavi, significatus* means *to give (facio) a sign (signum).* The ending *-ance* indicates that the word came from the noun *significantia*, formed by adding the feminine noun ending *-ia* to the *t*-stem adjective *significans, -antis.*

significant The Latin verb *significo, significare, significavi, significatus* means *to give (facio) a sign (signum).* The English word is the stem of its present participle *significans, significantis.*

significant digits The following story illustrates the ignorance of significant digits among the general population. The chemist Harold State once asked a guard at the Carnegie Mellon Museum how old the tyrannosaurus there was. "Five million and eight years old" was the answer. "How do you know that?" "When I was hired here eight years ago, they told me that it was five million years old." On another occasion, a student found the eccentricity of an elliptical plate at a local restaurant to fourteen decimal places after measuring the major and minor axes with a ruler. For the etymologies, see the entries **digit** and **significant**.

signum function The Latin noun *signum* means *sign*. This phrase is just a highfalutin name for the sign function.

similar The Latin adjective *similis* means *like, resembling*. The corresponding Greek adjective is ὅμοιος. The addition of the adjectival suffix -*aris* to the stem of what was already an adjective produced the late low Latin *similaris*, which entered French as *similaire*, whence came the English adjective. The low noun *similaritas* developed from *similaris*, and the French *similarité* proceeded from *similaritas*. The English *similarity* came from *similarité*.

similarity See the preceding entry.

similitude The Latin adjective *similis* means *like, resembling*. The related noun *similitudo, similitudinis* means *likeness, resemblance*.

simple The Latin adjective *simplex, simplicis* means *single, uncompounded, unmixed*.

simplex The Latin adjective *simplex, simplicis* means *plain, uncomplicated*. An *n-dimensional simplex* is an *n*-dimensional polytope that is the convex hull of its *n + 1* vertices.

simplicial This is a modern word produced by adding the Latin adjectival suffix -*alis* to the stem of what was already an adjective, to which an intermediate connecting vowel -*i*- had been appended.

simplify This verb is modeled on the analogy of verbs like *magnify*, as if there were a Latin verb *simplifacio, to make simple*, which there is not. The making of such -*fy* words is acceptable, provided that the stems to which the suffix is appended are of Latin origin.

simulation The Latin verb *simulo, simulare, simulavi, simulatus* means *to make like* and is connected with the adjective *similis*. The noun *simulatio, simulationis* is derived from the fourth principal part and means *the assumed appearance of anything*; its stem is the English noun.

simultaneous The Latin adverb *simul* means *at once, at the same time*, and is connected with the adjective *similis*. See the entry **instantaneous** above.

sine The Latin word *sinus, sinūs* means *a bending curve, a fold, a fold in a coastline, a bay, a gulf*. The use of this word for the trigonometric function is explained by D. E. Smith, Carl Boyer, and Dirk Struik as follows: The trigonometry of the mathematicians of the ancient world dealt with chords, not with ratios of lengths of sides of a triangle. The Sanskrit word for *chord*, according to one transliteration, is *jīva*. The Arabs merely transliterated this into their language. There is an Arabic word جيب meaning *bosom, bay, fold*, and this word displaced the transliteration, which must have been pronounced similarly to it. When Robert of Chester made his translations of Arabic mathematics into Latin, he translated جيب by *sinus*, whence we get our *sine*. The chord in a circle of radius r subtended by a central angle 2θ is of length $2r \sin \theta$. The quantity $r \sin \theta$ was originally called the *sinus rectus* of θ in order to distinguish it from $r(1 - \cos \theta)$, which was called the *sinus versus* or *versed sine of θ*. Euler was the first to take the radius of the circle always equal to unity. The *law of sines* in a triangle relates the lengths a and b of two sides to the angles α and β that they subtend: $(\sin a)/a = (\sin \beta)/b$.

single The Latin adjective *singulus* means *one at a time, one alone*; the syllable *sin-* is related to the syllable *sem-* of the adverb *semel*, which means *once*. Both *singularis* and *semel* are cognate with the Greek ἄμα, *at once*, as is the syllable *sim-* of *simul*, once.

singleton This is a strange word, the result of adding the suffix *-ton* to an adjective of Latin origin. It is an eighteenth-century invention on the analogy of *simpleton*, which existed in 1755 and was condemned by Dr. Johnson as low. The ending *-ton* is fanciful, without any etymological significance.

singular The Latin adjective *singularis, -e* means *alone, single, individual*. It is formed from the adjective *singulus, -a, -um, single, separate, one at a time, alone*, by the addition of the adjectival suffix *-aris*. See the entry **single**.

singularity The Latin noun *singularitas, singularitatis* means *the condition of being left alone, unity*. This became the French *singularité* and then the English *singularity*.

> If it is impossible to include some point in a circle of convergence of a power series representing the function $f(z)$, this point is called a singular point of the function. (Knopp, Part II, p. 82)

sinh This is the standard abbreviation for the hyperbolic sine function: $\sinh x = (e^x - e^{-x})/2$. The abbreviation stands for *sinus hyperbolicus*. The Latin word *sinus, sinūs* means *a bending curve, a fold, a fold in a coastline, a bay, a gulf*. The pronunciation *sinch* is comical. It should be read *hyperbolic sine*.

sinusoid This is a macaronic word. *Sinus* is mathematical Latin for *the sine function*, while *-oid* is from the Greek noun εἶδος, *form, shape, figure, that which is seen*, and, in the Aristotelian philosophy, the *species* as opposed to the *genus*. The word means *a function like the sine function*.

solid The Latin adjective *solidus, -a, -um* means *hard*. For centuries a *soldo* (plural *soldi*) was a unit of currency in Italy, *a solid [coin]*.

solution This is the stem of the Latin noun *solutio, solutionis*, formed from the fourth principal part of the verb *solvo, solvere, solvi, solutus*, which means *to loosen*.

solvable This is a French word taken over into English. A Latin adjective *solvibilis* would have been the natural ancestor, but the lexicons recognize no such word. *Solvable* is formed from the verb *solvo, solvere, solvi, solutus*, which means *to loosen*, and the adjectival suffix *-abilis*, so that the word naturally means *capable of being solved*. *Solvible* would have been the correct spelling since *solvo* is of the third conjugation.

space The Latin noun *spatium* means *space*.

species This is the Latin word meaning *a seeing, view, look* that was used for the Aristotelian technical term εἶδος. It is derived from the verb *specio, specere, spexi*, which means *to look at, behold*. It is related to the Greek verb σκέπτομαι of the same meaning.

specific This is a late and low Latin word formed by adding the suffix *-ficus* from *facio, to do*, to the stem of the verb *specio* and is intended to mean *making visible*.

spectral The Latin adjectival suffix *-alis* was added to the stem of the noun *spectrum, an image*, to produce the adjective *spectralis* with the meaning *pertaining to an image*.

spectrum This is a Latin noun meaning *an appearance, form, image of a thing*. It is derived from the verb *specto, spectare, spectavi, spectatus*, which is the frequentative of *specio, specere, spexi, to look at*.

sphere This is the Latin *sphaera*, which is the transliteration of the Greek σφαῖρα, which means *a ball*.

spherical The stem of the Latin adjectival suffix *-alis* was incorrectly superimposed upon the stem of the Greek adjectival suffix -ικός and the result was then added to the stem of a Greek noun to produce the English superadjective *spherical*. The correct word would have been s*pheric*, the transliteration, by way of the Latin *sphericus*, of the Greek adjective σφαιρικός, *ball-like*.

spheroid The Greek adjective σφαιροειδής means *like a ball*. It was transliterated into Latin as *sphaeroïdes*, of which *spheroid* is the English metamorphosis.

spiral The Latin noun *spira* means *anything coiled or wreathed*. It was natural that the adjectival suffix *-alis* should be added to its stem to form *spiralis* with the meaning *pertaining to a coil*. Since the Latin noun is the transliteration of the Greek σπεῖρα, an adjective *spiric* would have been correct.

sporadic The Greek verb σπείρω means *to sow, to scatter*. The associated adjective σποραδικός means *pertaining to scattering*, and was transliterated into Latin as *sporadicus*, whence is derived the English word.

square This is the Anglicization of a French corruption of the Latin prepositional phrase *ex quadra*, which means *from a carpenter's square*. The noun *quadra* is related to the numeral *quattuor, four*.

stable The Latin verb *sto, stare, steti, status* means *to stand*. The addition of the adjectival suffix *-abilis* to the stem of the second principal part produced the adjective *stabilis* with the meaning *firm, steady, able to stand on its own*.

standard This is the corruption of the French noun *étandard, a banner*, which appears in the Marseillaise, *Contre nous de la tyrannie, l'étandard sanglant est levé*. It is derived from the Latin *extendo, extendere, extendi, extensus*, which means *to stretch (tendo) out (ex), to unfurl*. It has nothing to do with the verb *to stand*.

stationary The Latin verb *sto, stare, steti, status* means *to stand*. From its fourth principal part was formed the noun *statio, stationis* with the meaning *a standing still, a stopping place*. To the stem of this noun was added the adjectival suffix *-arius* to produce *stationarius*, meaning *pertaining to stopping*.

statistical This is a modern word. The Latin verb *sto, stare, steti, status* means *to stand*. From the fourth principal part was formed the fourth-declension noun *status, status* with the meaning *condition, manner of standing*. In the Latin language, *status* did not mean *state*, which was *res publica*. To *status* there was first added the Greek nominal suffix of agent -ιστής. Then there was added the Greek adjectival suffix -ικός. To the concoction thereby created there was furthermore superimposed the Latin adjectival suffix *-alis* to produce *stat-ist-ic-al*, an extremely low word.

statistical inference See the entries **statistical** and **inference**.

statistics This is a very low word of the eighteenth century, the offspring of a succession of mistakes. The formula for it is *Latin status* + *Greek* -ιστής + *Greek* -ικός + *English s*. The name comes from the fact that the collection of data was originally an activity of the state. The first well-known example is in Luke II 1:

> Ἐγένετο δὲ ἐν ταῖς ἡμέραις ἐκείναις ἐξῆλθεν δόγμα παρὰ Καίσαρος Αὐγύστου ἀπογράφεσθαι πᾶσαν τὴν οἰκουμένην.

> Factum est autem in diebus illis, exiit edictum a Caesare Augusto ut describeretur universus orbis.

> And it came to pass in those days, that there went out a decree from Caesar Augustus, that all the world should be taxed.

The reference is to a registration of the inhabitants of the empire for the purpose of taxation.

stereographic The Greek adjective στερεός means *stiff*. The adjective γραφικός means *pertaining to writing* (γραφή). From these elements there was formed in the early nineteenth century the nice English word *stereographic*.

stochastic The Greek noun στόχος means *a target, a guess*. The associated verb στοχάζομαι means *to aim at, to guess*, and the noun στοχαστής means *a diviner, someone who guesses*. From this noun was formed the adjective στοχαστικός with the meaning *pertaining to guessing*.

stochastic discrimination This is the application of probability to the production of machines that read handwriting. Such machines are purchased by the U.S. Postal Service and many institutions that receive large amounts of handwritten mail. The State University of New York at Buffalo was an important center of such studies and the company Exegetics Incorporated of Blacksburg, Virginia, is prominent in the business.

strategy This is derived from the Greek word for *general*, στρατηγός. It is the abbreviation, by one letter, of the noun στρατηγία, *the office of a general.*

strict The Latin verb *stringo, stringere, strinxi, strictus* means *to draw tight together, to bind, tie.* The fourth principal part as an adjective means *close, tight.*

strophoid This modern word was formed by Isaac Barrow (1630–1677) from the Greek nouns στροφή, *a turning,* from στρέφω, *to twist,* and from εἶδος, *shape.* See the entry **-oid**. It was invented to be the name of the polar curve constructed in the following manner. Each non-vertical line ℓ through the point $A(a,0)$ intersects the y-axis at some point B. Let O be the origin. Find points P and P' on ℓ such that $BP = BP' = OB$. The locus of P and P' is the strophoid. The Cartesian equation is $y^2 = (a - x)x^2/(a + x)$. The polar equation is $r = a \cos 2\theta \sec \theta$. Like the *folium Cartesianum*, it has both loop and asymptote. The area of the region enclosed by the loop is $a^2(4 - \pi)/2$. The area of the region between the strophoid and its asymptote is $a^2(4 + \pi)/2$.

Smith's alternate definition of Barrow's strophoid is as follows. Consider the circle C with center $(a,0)$ and radius a, and let ℓ_1 be the line with equation $x = -a$, that is, $r = -a \sec \theta$. For each θ, $-\pi/2 < \theta < \pi/2$, consider the line ℓ_2 through the origin that makes an angle θ with the initial line. Then ℓ_2 will intersect C at some point Q (not the pole) and ℓ_1 at some point R. There is then a point P on ℓ_2 defined by $r = OP = OQ - OR = 2a \cos \theta - a \sin \theta$. The locus of such points P is the strophoid.

Barrow's strophoid is a special case of the following family of curves invented by J. Booth. Let ℓ_1 and ℓ_2 be two straight lines, each revolving with uniform speeds ω_1 and ω_2, respectively. Booth's strophoid is the locus of intersection of the two lines. Suppose, as we may, that at time $t = 0$, ℓ_1 has equation $y = (\tan \omega_1 t)x$ and ℓ_2 has equation $y = (\tan \omega_2 t)x + b$. If we put $c = \omega_2/\omega_1$, then the polar equation of Booth's strophoid is

$$r = b \cos c\theta \csc (1 - c)\theta.$$

If $c = 2$, this is Barrow's strophoid (rotated 90° counterclockwise).

student The Latin verb *studeo, studere, studui* means *to be eager, be earnest, take pains, strive after, be busy with*. The English noun is the stem of the present participle *studens, studentis*.

Student's t See the entries **student** and **t**. To preserve anonymity, W. S. Gosset (1876–1937) signed himself *Student* in his paper on this probability distribution.

sub *Sub* is the Latin preposition meaning *under*. The corresponding Greek preposition is ὑπό. Therefore *sub* should be used with Latin words and ὑπό with Greek words. To use *sub* with Greek words or ὑπό with Latin words is illiteracy. Such words sound and look ugly to those who know. With words of Germanic origin, the use of the prefix *sub-* is acceptable if the prefix *under-* is unpalatable. It is also correct to use the hyphen in words compounded with *sub* and *super*, such as *sub-ring* and *sub-group*, especially to avoid an unseemly concatenation of letters or to prevent mispronunciation due to an incorrect division of syllables.

subadditivity This is a property of measures. If $\{X, \mathbf{Q}, \mu\}$ is a measure space, then the property that $\mu(\cup A_n) \leq \Sigma\mu(A_n)$ for any sequence $\{A_n\}$ of measurable sets in $\mathbf{Q}$ is called *subadditivity*. The force of the prefix is a common one, *viz.*, that *equality* is replaced by *inequality* of the *less than or equal to* sort. See the entries **sub** and **additivity**.

subbase A Latin prefix has been added to a Greek noun, a mistake. An *open base* for a topological space $(X, \mathcal{J})$ is a family $\mathcal{J}'$ of open sets in

$\mathcal{J}$ with the property that every open set in $\mathcal{J}$ is the union of sets in $\mathcal{J}'$. An *open subbase* is a family $\mathcal{J}''$ of open sets in $\mathcal{J}$ such that the finite intersections of elements of $\mathcal{J}''$ are an open base for $(X, \mathcal{J})$. This use of the prefix *sub* is eccentric. See the entries **sub** and **base**.

subclass A subclass C' of a collection C of sets is any collection of sets such that $C' \subseteq C'$. The force of the prefix is a common one, *viz.*, that *set equality* is replaced by *set inequality* of the *included in but not necessarily equal to* sort. See the entries **sub** and **class**.

subcover If $\{X, \mathcal{J}\}$ is a topological space, a class $\{G_a\}$ of open subsets of X is an open cover for X if for all $x \in X$ there exists a $G_x \in \{G_a\}$ such that $x \in G_x$. If $\{G_a\}$ is an open cover for X, a class $\{G'_a\}$ of open subsets of X is an *open subcover* for X if $\{G'_a\}$ is an open cover for X and $\{G'_a\} \subseteq \{G_a\}$. The use of the prefix in this case is correct. See the entry **cover**.

subdiagonal A *subdiagonal entry* of a matrix $\{a_{i,j}\}$ is an entry $a_{i,i-1}$, that is, an entry directly underneath the diagonal. The force of the prefix here is that of *being below*. See the entries **sub** and **diagonal**.

subfactorial The *subfactorial function* is the integer-valued function f with domain the natural numbers whose value at any positive integer n is the number of permutations $a_1, a_2, \ldots, a_n$ of the first n positive integers $1, 2, \ldots, n$ such that $a_i \neq i$. One can easily calculate that $f(1) = 0$, $f(2) = 1$, $f(3) = 2$, etc. The use of the notation $!n$ for $f(n)$ is comical. The force of the prefix here is to replace *equality* by *inequality* of the *less than* sort, since $f(n) < n!$ See the entries **sub** and **factorial**.

subfield A field $(X', +', \times')$ is a *subfield* of a field $(X, +, \times)$ if $X' \subseteq X$ and $x +' y = x + y$ and $x \times' y = x \times y$ for all $x, y \in X'$. A Latin prefix has been added to a Germanic noun, a mistake. See the entry **sub**.

subgroup A group $(X', +')$ is a *subgroup* of a group $(X, +)$ if $X' \subseteq X$ and $x +' y = x + y$ for all $x, y \in X'$. A Latin prefix has been added to a Germanic noun, a mistake. See the entry **sub**.

subharmonic This is a low word, the marriage of the Latin *sub* and the Greek *harmonic*. It ought to have been *hypoharmonic*. See the entries **super** and **harmonic**.

submatrix A matrix A is a *submatrix* of a matrix B if A is obtained by deleting some rows and some columns from B. This is a daring use of the prefix *sub-*. See the entries **sub** and **matrix**.

submodule See the entries **sub** and **module**. Suppose M is a module whose set of scalars is the ring R. An additive subgroup A of the R-module M is a *submodule* of M if for all $r \in$ R and $x \in A$, $rx \in A$.

submultiple This noun is just a synonym of *factor*; if $x = ab$ where x, a, and b are integers, then a and b are submultiples of x. See the entries **sub** and **multiple**.

subnormal See the entries **subtangent**, **sub**, and **normal**.

subregion See the entries **sub** and **region**. A region A' that is a subset of a region A is called a *subregion* of A. This is a common and acceptable use of the prefix.

subring A ring $(X',+',\times')$ is a *subring* of a ring $(X,+,\times)$ if $X' \subseteq X$ and $x +' y = x + y$ and $x \times' y = x \times y$ for all $x, y \in X'$. See the entry **sub**.

subscript The Latin word *subscribo, subscribere, subscripsi, subscriptus* means *to write below, to sign one's name at the bottom of*.

subsequence A sequence $\{x_i\}$ is a *subsequence* of a sequence $\{y_i\}$ if $\{x_i\}$ is obtained from $\{y_i\}$ by deleting some members of $\{y_i\}$. See the entries **sub** and **sequence**.

subset A set A is a *subset* of a set B if $A \subseteq B$. A Latin prefix has been added to a Germanic noun, a mistake.

subspace See the entries **sub** and **space**. If $\{X, \varrho\}$ is a metric space and $Y \subseteq X$, then $\{Y, \varrho\}$ is the metric space produced by restricting ϱ to Y. $\{Y, \varrho\}$ is called a *subspace* of $\{X, \varrho\}$.

substitution The Latin verb *substituo, substituere, substitui, substitutus* means *to put (statuo) next, to put under (sub), to put in place of.*

subtangent Let C be a plane curve, $P(x,y)$ a point on C, and ℓ_1 and ℓ_2 the tangent line and normal line to C at P, respectively. Let A be the point $(x,0)$. Let ℓ_1 intersect the x-axis at T, and let ℓ_2 intersect the x-axis at N. Then $\|AT\|$ is the *subtangent* and $\|AN\|$ is the *subnormal* of C at P. The force of *sub* here is that of *under*. For the etymologies, see the entries **sub** and **tangent**.

subtend This is the stem of the Latin verb *subtendo, subtendere, subtendi, subtensus* which means *to stretch under*. It was formed to translate the Greek verb ὑποτείνω.

subtract The Latin verb *subtraho, subtrahere, subtraxi, subtractus* means *to draw up from below, to draw away secretly, to remove.*

subtraction This is the stem of the late Latin noun *subtractio, subtractionis*, which is formed by adding the nominal suffix *-io* to the stem of the fourth principal part of the verb *subtraho*. See the preceding entry.

subtrahend This is the stem of the gerundive *subtrahendus* of the Latin verb *subtraho*, which means *that which it is necessary to take away.* See the entry **subtract**.

success The Latin verb *succedo, succedere, successi, successus* means *to go (cedo) under (sub), to go from under, to ascend, to mount, to come after or into the place of.* The noun *successus, successūs* means *an advance uphill, approach, happy issue.*

succession Laplace's *law of succession* is the solution to the *sunrise problem*. If the sun has risen *n* times in succession from time immemorial through today, what is the probability that tomorrow morning it will rise again? Laplace's solution was *(n + 1)/(n + 2)*. That value is actually the mean of the random variable with beta distribution with parameters α = *n + 1* and β = *1*. An equivalent formulation was given by Clyde Haberman in the November 1, 1981, issue of *The New York Times*: The reelection of Mayor Koch is as certain as the sunrise.

successive The Latin verb *succedo, succedere, successi, successus* means *to go (cedo) under (sub), to go from under, to ascend, to mount, to come after or into the place of*. Evidently the participle *succedens, succedentis* was felt in medieval times inadequate to express the idea of coming after, so the adjectival suffix *-ivus* was added to the stem of *successus* to produce the adjective *successivus*.

sufficient The Latin verb *sufficio, sufficere, suffeci, suffectus* means *to put under, to provide, supply, be adequate*. It is compounded of the verb *facio, to do*, and the preposition *sub, under*. The present participle is *sufficiens, sufficientis*, and the English word is the stem of the Latin participle.

sum This word is an abbreviation of the Latin noun *summa* derived from the superlative adjective *summus, -a, -um*, which means *the highest*.

summa See the previous entry.

summable This word means *able to be summed* and is derived by the addition of the suffix *-able* from the Latin *-abilis* to the stem of the medieval Latin verb *summo, summare, to add up*.

summation This word is the creation of modern mathematicians writing in Latin, who derived *summatio, summationis* in a natural way from the fourth principal part of the medieval verb *summo, summare, summavi, summatus*. The way to say *to sum* in proper Latin was *summam facere* or *summam computare*.

super This prefix is the Latin preposition *super*, which means *above*. The Greek equivalent is ὑπέρ.

superadditive This is the composition of the Latin prefix *super-* and the adjective *additivus*. See the entries **super** and **additive**.

superdiagonal A *superdiagonal entry* of a matrix $\{a_{ij}\}$ is an entry $a_{i,i+1}$. See the entries **super** and **diagonal**.

superharmonic This is a low word, the marriage of the Latin *super* and the Greek *harmonic*. It ought to have been *hyperharmonic*. See the entries **super** and **harmonic**.

superimpose This verb is composed of the prefix *super-*, *above*, and the verb *impose*, which is derived from the fourth principal part of the Latin verb *impono, imponere, imposui, impositus*, which means *to put (pono) on (im-* from *in*).

superior This is the Latin adjective meaning *higher*, the comparative degree of the adjective *superus*, which means *situated above*.

superpose The Latin verb *superpono, superponere, superposui, superpositus* means *to place (pono) over or upon (super)*. The English word is the corruption of the fourth principal part. It is a correct alternative to *superimpose*.

superposition The noun *superpositio, superpositionis* is formed in a natural way from the verb *superpono* (see the previous entry), but its meaning in Roman times was a *paroxysm of a disease*. Its use as a mathematical technical term is modern. *Superposition* is a method of proof utilized sparingly by Euclid in the *Elements*, whereby he permitted himself to move a geometrical figure through space without distorting it in order to place it upon another figure. Other geometers, like Heron, gave alternate proofs by superposition for propositions (for example, I 5) proved by Euclid by contradiction, which they disliked.

superscript This is the stem of the fourth principal part of the Latin verb *superscribo, superscribere, superscripsi, superscriptus*, which means *to write (scribo) above (super)*.

superset This is a low word, the result of imposing a Latin prefix on an Anglo-Saxon noun. If *A* is a subset of *B*, then *B* is a *superset* of *A*.

supplement This is the stem of the Latin noun *supplementum*, which means *a filling up*, from the verb *suppleo, supplere, supplevi, suppletus, to fill up, to make complete. Suppleo* itself is derived from the preposition *sub, under*, and the verb *pleo, to fill*.

support This is the stem of the first principal part of the Latin verb *supporto, supportare, supportavi, supportatus*, which means *to bear (porto) from below (sub), to carry up*.

supremum This word is the nominative neuter singular of the Latin superlative adjective *supremus, -a, -um* from which we get our word *supreme*. The positive degree is *superus*; the comparative degree is *superior*.

surd This noun is the stem of the Latin adjective *surdus*, which means *deaf*. The Greek word for *irrational* was ἄλογος, which, in addition to its transferred, mathematical sense of *not in proportion and therefore not reasonable*, had the literal meaning *not having the use of words* (λόγοι). Since the deaf are people unable to speak correctly, the Arabs translated ἄλογος by ܐܨܡ, which means *deaf*, which was in turn translated literally into Latin by *surdus*. Thus, the word *surd* entered into the mathematical vocabulary as a result of a confusion of the literal and transferred meanings of the word ἄλογος.

surface This word is a corruption of the Latin noun *superficies*. The Latin prefix *super-* developed into *sur-* in France. *Face* is derived from the Latin *facies*.

surjective The Latin verb *superiacio, superiacere, superieci, superiectus* means *to throw (iacio) over (super)*. The Latin prefix *super-* developed into *sur-* in France. This word has no existence except as a mathematical technical term.

survey The medieval Latin verb *supervideo, supervidere, supervidi, supervisus* meaning *to look upon* developed into the French *survoir*, the proximate ancestor of *survey*. The fourth principal part of the Latin verb produced our verb *supervise*.

syl-, sym-, syn-, sys- This prefix is derived from the Greek preposition σύν, which means *with*. It should therefore only be attached to words of Greek origin.

syllogism This word is the stem of the Greek noun συλλογισμός, which means *a reckoning all together, a reckoning up, a collecting from premises*. It is derived from the verb συλλογίζομαι, *to reckon all together*.

symbol This word is the stem of the Greek noun σύμβολον, *mark, sign*, from the verb συμβάλλω, *to bring or throw* (βάλλω) *together* (σύν). The Greek noun later developed the meaning of *Creed*, as in the *Nicene Creed*.

symmetric The word is the stem of the make-believe Greek adjective συμμετρικός, which was imagined to mean *of like measure or size with*, but the actual Greek adjective with that meaning is σύμμετρος. The *symmetric difference* of two sets, so-called because the order in which one gives the sets is not important, is the set of elements that are in exactly one of the two sets.

symmetry The Greek adjective σύμμετρος means *measured or commensurate with, of like measure or size with*, from σύν, *with*, and μέτρον, *a measure*. From the adjective is formed the noun συμμετρία, which means *due proportion*. Greek nouns ending in -ία come into English ending in -*y*.

symplectic The Greek verb συμπλέκω means *to twist* (πλέκω) *together* (σύν), and the associated adjective συμπλεκτός means *twined together*. This word is the Greek equivalent of the Latin *complex*.

synthesis The Greek noun σύνθεσις means *a putting together*. It is derived from the verb συντίθημι, *to put together*. It is a name for the standard direct type of proof, wherein one proceeds from the hypothesis to the conclusion. It is opposed to *analysis*, *q.v.*

synthetic This is the stem of the Greek adjective συνθετικός, which means *skilled in putting things together*, from the verb συντίθημι, *to put* (τίθημι) *together* (σύν).

system This is the stem of the Greek noun σύστημα, which means *that which is put together, a composite whole*. The corresponding verb is συνίστημι, *to stand* (ἵστημι) *together* (σύν).

T

t The Latin *t* is the Greek τ, *tau*. The Greek τ is used in mathematics to represent the torsion of a space curve at a point. The corresponding Hebrew letter *taw* ת was used by Cantor to describe a certain collection of cardinal numbers; his use of Hebrew letters was a bad idea, as one can scarcely expect people to write legibly in their own language, let alone in Hebrew. In topology there are T_1, T_2, T_3, and T_4 spaces, too many to be remembered accurately. If $\{X, \mathcal{J}\}$ is a topological space, $\{X, \mathcal{J}\}$ is a T_1 space if given two distinct points x and y in X, there is an open set in $\mathcal{J}$ which contains y but not x. T_2, T_3, and T_4 spaces are just other names for Hausdorf space, regular space, and normal space, respectively.

table The Latin word for *table* is *tabula*. The word *table* came directly into English from French.

tabular *Tabula* is the Latin word for *table*. The addition of the suffix *-aris* to the stem produces the adjective *tabularis, pertaining to a table*. The suffix *-aris* was used instead of the suffix *-alis* for the sake of euphony.

tangency See the entry **tangent**. The noun *tangency* is formed as if from a Latin noun *tangentia*, but such a noun is not known to have existed.

tangential See the entry **tangent**. The Latin adjectival suffix *-alis* was added to the stem of the participle *tangens, tangentis*, which was already an adjective but felt to be a noun, *the tangent*.

tangent This is the stem of the Latin *tangens, tangentis,* the present participal of the verb *tango, tangere, tetigi, tactus,* which means *to touch*.

tanh This is the standard abbreviation for the hyperbolic tangent function, defined by *tanh x = (sinh x)/cosh x*. The abbreviation stands for *tangens hyperbolica*. If a body of mass m falls from rest from a height above the earth, if the only forces acting are vertical, *viz.,* that of gravity and that of friction, and if the force due to friction at any moment is directly proportional to the square of the velocity v at that moment, then the solution to the resulting differential equation is

$$v(t) = (mg/k)^{1/2} \, tanh[(kg/m)^{1/2}t],$$

where t is time, g is the acceleration of gravity, and k is the constant of proportionality.

tautochrone The ungrammatical juxtaposition of the Greek words ταὐτό, *the same,* and χρόνος, *time*; it is supposed to be ταὐτόχρονος and mean *the curve of same times*, that is, the plane curve along which a point mass falling under gravity and without friction will reach the bottom in the same time no matter from which higher-up point it starts from rest. That plane curve is the cycloid. Huygens (1629–1695) proved that the same time is always $\pi(r/g)^{1/2}$, where r is the radius of the rolling circle that produced the cycloid in question and *g* is the acceleration of gravity. Whoever coined it (its first appearance in print, according to the *Oxford English Dictionary*, is in the French *Dictionnaire Trévoux* of 1771) must have thought that ταὐτό was an adjective, whereas it is really a noun. In properly formed Greek compounds beginning with ταὐτό, the second part must come from a verb of which *the same* is the object. For example, a *tautologous* statement is a statement that says something that has already been said. If a Greek word meaning *the same time* were to be created, it would have to be αὐτόχρονος. Though sanctioned by immemorial custom, it is better to put it on the back burner and use the word *isochrone*, which is formed correctly. In favor of *tautochrone* is that its Latin equivalent was used by Euler, *tautochronus, -a, -um.*

tautology As pointed out in the preceding entry, in Greek a ταὐτολόγος is a person who says what has already been said, that is, who says (λέγω) the same (ταὐτό). From this good word later ignorant authorities formed the noun ταὐτολογία, *tautology*, which is supposed to mean *something that has already been said*. It does not follow the proper rules for noun construction in Greek, but it is now acceptable because of immemorial custom.

tensor The Latin verb *tendo, tendere, tetendi, tensus* means *to stretch*; it is related to the Greek verb τείνω with the same meaning. The noun *tensor* is therefore *someone who stretches*. The tensor product $\mathcal{U} \otimes \mathcal{V}$ of two finite dimensional vector spaces $\mathcal{U}$ and $\mathcal{V}$ over the same field is defined to be the set of mappings $x \otimes y, x \in \mathcal{U}, y \in \mathcal{V}$, whose domain is the set of bilinear forms on $\mathcal{U} \oplus \mathcal{V}$ and whose value at such a form f is given by $(x \otimes y)(f) = f(x,y)$.

tenure This is the medieval Latin *tenitura, a holding*, from the verb *teneo, tenere, tenui, tentus*, which means *to hold*.

tera- The Greek noun τέρας, τέρατος means *a sign, wonder, marvel*. This prefix has comically been assigned to be the name of the number *trillion*, that is, of 10^{12}. This convention is absurd because there is no reason why *tera* should mean 10^{12}; why not 10^{15} or 10^{18}? There is no connection between a portent in the heavens and the number *twelve*.

term This word is an abbreviation of the stem of the Latin noun *terminus*, which means a *boundary*. The noun *terminus* was the standard Latin translation of the Greek mathematical term ὅρος, *boundary, end*, found in the definitions of Euclid's *Elements*.

terminal *Terminus* in Latin is *the end* or *the boundary*. The addition of the suffix turns the word into the adjective *terminalis, pertaining to the end*.

ternary The Latin distributive numeral *terni, -ae, -a* means *three at a time*. The addition of the adjectival suffix *-arius* at the end of the stem produced the adjective *terminarius* meaning *containing or consisting of three*.

tessellation The Latin noun *tessera* is "a cube of wood, stone, or other substance, used for various purposes" (Cassell); in modern Italian it means *a pass* to enter a museum. It is related to the Greek number τέσσαρες, *four*. Its diminutive is *tessella*, a small *tessera, a small square piece of stone*. From this noun proceeded the denominative verb *tessello, tessellare, tessellavi, tessellatus, to set with stones*, whose fourth principal part *tessellatus* means *set with small stones, mosaic*. The English *tessellation* is constructed as if from a Latin noun *tessellatio, tessellationis*, but no such noun is known.

test The Latin verb *torreo, torrere, torrui, tostus* means *to burn or parch*. From it was derived the noun *testa* meaning *a piece of burnt clay, a pot, a tile*. Weekeley writes *sub voce* that a *test* was originally an alchemist's cupel, used in assaying gold and silver, from which employment are derived the expressions *to bring to the test, to stand the test*. The verb *to test* he condemns as an Americanism and says that it is first recorded to have been used by George Washington.

tetracuspid This is supposed to mean a curve with four cusps, but the proper word would have been *quadricuspid*. As it stands, it is a macaronic word, *tetra* coming from the Greek word τέτταρες for *four* and *cuspid* from the Latin word *cuspis, cuspidis,* which means *the point,* especially *of a spear*.

tetrahedron This is a figure with four "seats." It is the Greek τετράεδρον. The Greek word is actually the neuter form of the adjective τετράεδρος, *having four sides*.

text The Latin verb *texo, texere, texui, textus* means *to weave, to twine together, to put together, to compose*. The related noun *textum* means *woven cloth, web, fabric, the style of a written composition*. The fourth-declension noun *textus* was used in medieval times to refer to the *ipsissima verba* of the Bible. Weekeley writes that "A *text-book* was originally a classic written wide to allow of interlinear glosses."

theorem The Greek verb θεωρέω means *to look at, view, behold, consider,* whence comes the noun θεώρημα, *a sight, a spectacle, a thing contemplated by the mind*. Euclid ended certain of his propositions with the words, ὃ ἔδει δεῖξαι, "This is what it was necessary to prove." These were called by the commentators *theorems*. The remainder he ended with the words ὃ ἔδει ποιῆσαι, "This is what we wanted to do." These were called by the commentators *constructions*. Euclid himself did not use the word *theorem*; his propositions were numbered, but he did not refer to those numbers in later demonstrations. Statements like "By proposition 1 of Book I" were added by later translators and commentators.

theory The Greek verb θεωρέω means *to look at, view, behold, consider,* whence comes the noun θεωρία, *a beholding, contemplation*.

theta The capital letter is Θ; the small case letter is θ. The use of the latter for the former is common and a sign of illiteracy.

Timaeus *Timaeus* is a Greek proper name, the name of the astronomer and mathematician of the fourth and fifth centuries B.C.

who is the eponymous main speaker of the dialogue of Plato. The name is derived from τιμή, the Greek word for *honor*.

> Adjectives signifying *belonging* or *related* in any way *to* a person or thing are formed from noun stems by the suffix -ιο (nom. ιος). (William W. Goodwin, *A Greek Grammar*, St Martin's Press, New York, 1968, §850)

Thus we have τιμα + ιος = Τίμαιος, transliterated into the Latin *Timaeus*, the hero of the dialogue, *the honorable one*.

In the dialogue *Timaeus*, Plato taught the doctrine that the world was created by a mathematical god, an idea that had a profound impact on the study of mathematics, which was seen as a result to be a religious activity. Bertrand Russell summarized and evaluated Plato's *Timaeus* in the chapter *Plato's Cosmogony* of his *History of Western Philosophy*:

> [The *Timaeus*] had more influence than anything else in Plato, which is curious, as it certainly contains more that is simply silly than is to be found in his other writings....It is difficult to know what to take seriously in the *Timaeus*, and what to regard as play of fancy. (Bertrand Russell, *History of Western Philosophy and Its Connection with Political and Social Circumstances from the Earliest Times to the Present Day*, Folio Society, London, 2004, pp. 139, 144)

The best commentary on this dialogue is by A. E. Taylor, *A Commentary on Plato's Timaeus*, Oxford, at the Clarendon Press, 1928.

The action in the dialogue *Timaeus* takes place on a holiday, the feast of the goddess Bendis, the day after the conversations recorded in the *Republic*. In the *Republic*, Socrates had discoursed on the state—how constituted and of what citizens composed it would seem likely to be most perfect. Three of the people who had heard him speak before, Timaeus the mathematician, Critias, and Hermocrates, now come to visit Socrates to fulfill a pledge that they had made to him the evening before, to describe in action, in a sort of novel, the state whose constitution he had outlined in the *Republic*. Critias reports that he can do no better than present a mere novel; indeed, his great-grandfather had heard from Solon that the Athenians had long ago had a state like the one outlined by Socrates

314

in the *Republic*, and he was ready to relate Solon's unfinished poem on the subject.

Solon, it appears, had gone to Egypt to consult with the priests there about antiquity, and he related to them the various Greek myths about what had gone on in those times. The Egyptians laughed at him and said that they had much more ancient stories. The reason for this, they said, was that the earth had many times been destroyed by flood and fire, and the Greeks remembered only the most recent catastrophe since, in each destruction, records of the previous destruction had been wiped out. The Egyptians, however, had escaped these calamities alone of all peoples because the Nile saves them when the rest of the earth goes up in flames; furthermore, since it never rains in Egypt, when the rest of the world is drowned in water, the Egyptians escape unharmed. The priests told Solon that before the last deluge, around 9600 B.C., the Athenians had the best governed of all cities, and had saved the whole world from the invasion of the vast empire of Atlantis, which was attempting to conquer the world at that time. After the defeat of Atlantis at the hands of the Athenians, the great deluge occurred; the record of the valor of the Athenians was wiped out with everything else, including the island of Atlantis, which went to the bottom of the ocean. For, accompanying the deluge there was a great earthquake, and Atlantis was covered up by the resulting tidal wave.

Critias now proposes that he present a novel with these ancient Athenian heroes as his characters; their state was surely most like the ideal one outlined by Socrates in the *Republic*. But before doing so, he suggests that an account first be given of the creation of the world. He says, "Our intention is, that Timaeus, who is the best astronomer among us, and has made the nature of the universe his special study, should speak first, beginning with the generation of the world and going down to the creation of man" (27 a 3–6). Critias will then take over and discuss the old Athenian state. The dialogue *Timaeus* is thus the middle work of a trilogy whose first and last parts are the *Republic* and *Critias*, respectively. The trilogy was never completed because Plato died while composing the *Critias*, right in the middle of a sentence. As Cicero reported, *Plato uno et octagesimo anno scribens est mortuus,* Plato died while writing at eighty-one years of age (*De Senectuste* V 13).

The great cosmological myth begins at this point.

> In the *Timaeus,* which is Plato's "monograph" on cosmogony and cosmology, there is a meaningful mixture of mathematics and myth-making. (Solomon Bochner, *The Role of Mathematics in the Rise of Science,* p. 16)

When we think of the word *myth,* we think of a silly fable that not even the credulous believe nowadays. This was not the case in Plato's time. As Bishop Westcott pointed out in his essay on the subject,

> The Platonic myth is, in short, a possible material representation of a speculative doctrine, which is affirmed by instinct, but not capable of being established by a scientific process....There are two great problems with which the Platonic myths deal, the origin and destiny of the Cosmos, and the origin and destiny of man. ("The Myths of Plato," in *Essays in the History of Religious Thought in the West,* Macmillan and Company, London, 1891, pp. 6, 11)

We may best compare them with the stories in the Bible, which are believed to be literally true by many, and have been studied by most of the rest. Because of these myths and speculations, Plato became the father of many heresies, and the medieval Church chose the philosophy of Aristotle as the one that was most appropriate to be the handmaiden of theology. Nevertheless, Christianity always had famous Platonists in its ranks, who held the view that John Addington Symonds attributed to Marsilio Ficino:

> He maintained that the Platonic doctrine was providentially made to harmonize with Christianity, in order that by its means speculative intellects might be led to Christ. ("Marsilio Ficino," in *Encyclopaedia Britannica,* 11th edition, vol. x, p. 318a)

Timaeus begins his account by saying that the universe is physical, and therefore was created. Its creator, he says, was good, and "desired that all things should be as like himself as they could be" (29 e 3). "...God desired that all things should be good and nothing bad, so far as this was attainable" (30 a 2–3). The perfection of the universe required that it be unique. So, out of the primeval chaos, he created the universe as we now know it, which is as good as it could be, as Leibnitz and Rousseau were to say later. Plato will have nothing of Tennyson, who spoke of "nature red in tooth and claw" (*In Memoriam: Arthur Henry Hallam*, MDCCCXXXIII, ix 15). The world was created a living being with a soul and intelligence, a perfect and unique animal. This teaching that the universe is an animal has been resurrected even in our modern age as the *Gaia theory*, of which we read in the August 29, 1989, issue of *The New York Times*, and according to which the earth, at any rate, is a living organism.

What, then, was the plan according to which the creator fashioned the universe? First of all, there had to be four elements, just as there are four seasons, four points of the compass, and four gospels. The visibility of the world required that there be fire; its solidity required that there be earth. The existence of the other two elements is explained by recourse to the fact that between two perfect cubes e^3 and f^3 (for each element is three-dimensional) one can always insert two whole number proportionals f^2e and fe^2:

$$f^3/f^2e \;=\; f^2e/fe^2 \;=\; fe^2/e^3;$$

there could not be just three elements because a single proportional inserted between f^3 and e^3 would usually be irrational:

$$f^3/x \;=\; x/e^3 \;\Rightarrow\; x = (f^3e^3)^{1/2} \text{ (in most cases irrational).}$$

Furthermore, God created the world a sphere because that is the surface of constant curvature, and the world animal must be everywhere the same on the outside because, unlike other animals, it has no need of eyes, ears, or limbs since there is nothing outside of itself for it to see, hear, or move towards.

The soul of the universe, which we may think of as the laws that govern its activities, the creator made as follows. He first considered the two geometric progressions

$$1 \quad 2 \quad 4 \quad 8 \qquad \text{and} \qquad 1 \quad 3 \quad 9 \quad 27,$$

which are customarily exhibited in the form of a *lambda*:

$$
\begin{array}{ccccc}
& & 1 & & \\
& 2 & & 3 & \\
4 & & & & 9 \\
8 & & & & 27
\end{array}
$$

When combined, these are meant, it seems, to give the relative distances from the moveable stars to the earth. (Plato's system was geocentric.)

1(Moon) 2(Sun) 3(Venus) 4(Mercury) 8(Mars) 9(Jupiter) 27(Saturn)

He then inserts the arithmetic and harmonic means between successive numbers in each of the two sequences just enumerated; recall that the harmonic mean of a and b is the number x defined by

$$a/b = (x-a)/(b-x) \quad or \quad x = 2/[(1/a) + (1/b)],$$

so that

$$x = 2ab/(a + b).$$

We then get the fuller sequences

$$1 \quad 4/3 \quad 3/2 \quad 2 \quad 8/3 \quad 3 \quad 4 \quad 16/3 \quad 6 \quad 8$$

$$1 \quad 3/2 \quad 2 \quad 3 \quad 9/2 \quad 6 \quad 9 \quad 27/2 \quad 18 \quad 27.$$

We observe that the successive ratios are all 4/3, 3/2, or 9/8; "these are the ratios which correspond to the melodic intervals of the major fourth, major fifth, and major tone" (Taylor, *A Commentary on Plato's Timaeus*, p. 139), so it is clear that the creator is aiming at the creation of a musical scale, which the Greeks called a *harmonia*. Plato is teaching that the harmony and order of the universe depend on ratios of natural numbers corresponding to the consonant intervals of a musical scale; to him is due the loathing and incomprehension of *irrational* (that is, unreasonable) numbers. By combining the two expanded sequences, we produce the great sequence

1 4/3 3/2 2 8/3 3 4 9/2 16/3 6 8 9 27/2 18 27.

If one had fifteen strings whose lengths were in the ratios of the numbers in the great sequence, they would make, when plucked, the musical sounds of the scale. (Plato does not discuss the problem that by combining the two sequences, we produce successive terms 9/2 and 16/3 whose ratio is not 4/3, 3/2, or 9/8.) In the story of the Myth of Er in the tenth book of the *Republic*, Plato has put a Siren on each planet or star to emit the appropriate sound; the Middle Ages replaced these with angels.

> There's not the smallest orb that thou behold'st
> But in his motion like an angel sings,
> Still quiring to the young-eyed cherubins;
> Such harmony is in immortal souls;
> But whilst this muddy vesture of decay
> Doth grossly close it in, we cannot hear it.
> (Shakespeare, *Merchant of Venice* V 1)

> But else in deep of night, when drowsiness
> Hath locked up mortal sense, then listen we
> To the celestial Sirens' harmony
> That sit upon the nine enfolded spheres,
> And sing to those that hold the vital shears
> And turn the adamantine spindle round
> Of which the fate of gods and men is wound.
> Such sweet compulsion doth in music lie,
> To lull the daughters of necessity,
> And keep unsteady Nature to her law,
> And the low world in measured motion draw
> And the heavenly tune, which none can hear
> Of human mold with gross unpurged ear.
> (Milton, *Arcades*, the speech of Genius)

Writers after Plato claimed to know all about this "Music of the Spheres," and a choir of eight of them would have had no difficulty (so James Adam claimed, in his commentary on Plato's *Republic*, vol. II, p. 453) in rendering it on a small scale. Succeeding generations were most fascinated by the subject, and the great Kepler wrote a huge work on it, *Five Books on the Scales of the World*, but his exhaustive researches on the subject are now read with derision or just ignored, except for the third law of planetary motion, which he discovered during his attempts to get the right notes, and which is the only part of the immense volume of any scientific value.

> His observations on the three comets of 1618 were published in *De Cometis*, contemporaneously with *De Harmonice Mundi* (Augsburg, 1619), of which the first lineaments had been traced twenty years previously at Gratz. This extraordinary production is memorable as having announced the discovery of the "third law"—that of the sesquiplicate ratio between the planetary periods and distances. But the main purport of the treatise was the exposition of an elaborate system of celestial harmonies depending on the various and varying velocities of the several planets, of which the sentient soul animating the sun was the sole auditor. The work exhibiting this fantastic emulation of extravagance with genius was dedicated to James I of England, and the compliment was acknowledged with an invitation to that island, conveyed through Sir Henry Wotton. (Agnes Mary Clerke, "Kepler," in *Encyclopaedia Britannica*, eleventh edition, vol. 15, p. 750b)

In our own time, Paul Hindemith gave a musical interpretation of the Music of the Spheres in his opera *The Harmony of the Universe* (1957), which is based on Kepler's life.

The intention of the demiurge (Plato's name for the creator) was to make the world as perfect as possible by using art and reason, that is, mathematics.

> Having these purposes in mind, he created the world a blessed god (34 b 8–9)....When the father and creator saw the creature, which he had made, moving and living, the created image of the eternal gods, he rejoiced. (37 c 6–7)

He then made a moving image of eternity, an image that he called *time*.

After creating the universe, the creator next made the immortal gods. He then created a large number of souls, which he placed in the stars, one soul in each star, and he ordered the gods to make the bodies into which these souls would go, the bodies of birds, the bodies of animals, including man, and the bodies of fish. According to Plato, therefore, the souls are all pre-existent, created at one time at the beginning of the world, a very great heresy indeed. Before the souls went into the bodies, the creator showed them, from their stars, the universe he had created and explained to them the mathematical laws according to which it functioned. The souls were then sent forth to live in the bodies of men, which had been fashioned by the gods.

> He who lived well during his appointed time was to return and dwell in his native star, and there he would have a blessed and congenial existence. But if he failed in attaining this, at the second birth he would pass into a woman (42 b 1–c 1)...for of the men who came into the world, those who were cowards or led unrighteous lives may with reason be supposed to have changed into the nature of women in the second generation (90 e 6–91 a 1)....But the race of birds was created out of light-minded men, who although their minds were directed toward heaven, imagined, in their simplicity, that the clearest demonstration of the things above was to be obtained by sight; these were remodeled and transformed into birds, and they grew feathers instead of hair. The race of wild animals, again, came from those who had no philosophy in any of their thoughts, and never considered at all about the nature of the heavens, because they had ceased to use their heads, but followed instead their breasts. In consequence of these habits of theirs, they had their front legs and their heads resting upon the earth to which they were drawn by natural affinity, and the crowns of their heads were elongated and of all sorts of shapes, into which the courses of the souls were crushed by reason of disuse. And this was the reason why they were created quadrupeds and polypods: God gave the more senseless of them the more support that they might be more attracted to the earth. And the most foolish of them, who trail their bodies entirely upon the ground and have no longer any need of feet, he made without feet to crawl upon the earth. The fourth class were the inhabitants of the water; these were made out of the most entirely senseless and ignorant of all, whom the transformers did not think any longer worthy of pure respiration, because they possessed a soul which was made impure by all sorts of transgressions; and instead of the subtle

and pure medium of air, they gave them the deep and muddy sea to be their element of respiration; and hence arose the race of fishes and oysters, and other aquatic animals, which have received the most remote habitations as a punishment of their outlandish ignorance. These are the laws by which men and animals pass into one another, now, as ever, changing as they lose or gain wisdom and folly. (91 d 6–92 c 3)

The order of condemnation of the wicked soul is therefore from man to woman to bird to beast to fish, and the order of salvation is the opposite, from fish to beast to bird to woman to man. The life of anything other than that of a just male is a sort of purgatory. The soul can return to its star only from the body of a man, not from the body of any other; the soul of a woman or an animal can never find release unless it comes into the body of a just man; man is the redeemer of nature. The singing of the birds is mourning, for they yearn to return to the stars, whence their souls came. For Plato, it is the soul that is the individual; the body is quite secondary, and the soul may go through many, many bodies before achieving salvation. The passing back and forth between animals and men was found quite unseemly by Aristotle, who would not allow it, and the Christians threw the whole thing out because for them the principle of individuation had to be the body, or their dogma of the bodily resurrection would fall. The Platonic teaching of the creation of all the souls at the beginning of the world and of their transmigration into many metamorphoses (metempsychosis) was finally condemned by the emperor Justinian and the Fifth Ecumenical Council (Constantinople II) in 553 (Denzinger, *Enchiridion Symbolorum et Definitionum*, ed. vii, §187) and replaced by the doctrine that the soul was created at the same time as the body and was for that body only, and, indeed, only for men's bodies; animals were wickedly and unetymologically said not to have souls, although the name *animal* itself means a thing that has a soul. Later Thomas Aquinas and Dante condemned the Platonic heresy.

[Consider] the opinion of certain philosophers of old who maintained that the souls return to the stars that are their compeers. But this is absolutely absurd....The soul...cannot pass from one body to another. (*Summa Theologiae*, Tertia, Suppl., Quaest. xcvii, Art. 5)

Ancor di dubitar ti dà cagione
Parer tornarsi l'anime a le stelle
Secondo la sentenza di Platone...
Quel che Timeo de l'anime argomenta
Non è simile a ciò quel che si vede,
Però che, come dice, par che senta.
(*Divina Commedia, Paradiso*, IV, 22–24, 49–51)

Another thing that gives you cause for doubt
Is the doctrine taught by Plato
That the souls return to the stars...
That which Timaeus teaches about the souls
(And it appears that he believes that which he says)
Is not like what you see here.

Having described the creation of the universe and the doctrine of reincarnation, Plato next takes up the structure of the four elements: fire, earth, air, and water.

> In the first place, then, as is evident to all, fire and earth and water and air are bodies. And every sort of body possesses solidity, and every solid must necessarily be contained in planes; and every plane rectilinear figure is composed of triangles; and all triangles are originally of two kinds (τὰ δὲ ἐτρίγωνα πάντα ἐκ δυοῖν ἄρχεται τριγώνοιν), both of which are made up of one right and two acute angles; one of them has at either end of the base the half of a divided right angle, having equal sides, while in the other the right angle is divided into unequal parts having unequal sides (53 c 4–d 4)....Now of the two right triangles, the isosceles right triangle has one form only, but the scalene or unequal-sided triangle has an infinite number of forms (54 a 1–2)....Now the one which we maintain to be the most beautiful of all these unequal-sided triangles is that of which the double forms a third triangle which is equilateral. (54 a 5–7)

The basic building blocks of the four elements are therefore the isosceles right triangle and the 30-60-90 right triangle.

> And next we have to determine what are the four most beautiful bodies which are unlike one another, and of which some are capable of resolution into one another. (53 d 7–e 2)

These four bodies, Plato says, are the tetrahedron, the cube, the octahedron, and the icosahedron. The tetrahedron, the octahedron, and the icosahedron have for their faces four, eight, and twenty equilateral triangles, respectively, and each of these equilateral triangles can be divided down the middle into two 30-60-90 right triangles. The odd one, the cube, has six squares for its faces, and each of these squares can be divided along the diagonal into two isosceles right triangles. Each element is assigned to its proper solid by the following reasoning. Since fire is the most acute of the elements, it must be made up of tetrahedra, the tetrahedron having the sharpest angles. Since earth is the most stable and immoveable of the elements, it must be made up of cubes since the cube is the most stable of the four bodies. Since water flows so freely, it must be made up of the most moveable bodies, *viz.*, the icosahedra. By default, air is made up of octahedra. Timaeus next points out that since these atoms are so small, one cannot make them out with the naked eye.

> We must imagine these to be so small that no single particle of
> any of the four kinds is seen by us on account of their smallness,
> but when many are collected together, their aggregates are seen.
> (56 b 7–c 3)

There is a fifth body, the dodecahedron, related to the four that we have just mentioned, but Plato could not use it because the twelve faces are pentagons, and pentagons cannot be resolved into 30-60-90 and isosceles right triangles because each of their interior angles is 108°. Instead, he says, "There was yet a fifth body which God used in the delineation of the universe" (55 c 4–6). The reason why God used the dodecahedron when he created the universe is that the universe was to have a spherical shape, and of the five "Platonic" solids, the tetrahedron, the cube, the octahedron, the dodecahedron, and the icosahedron, the icosahedron is the one which, when inscribed in a sphere, takes up the greatest percentage of the volume, as the following table shows.

Solid	Fraction of the volume of the circumscribed sphere that the Platonic solid occupies		
Tetrahedron	$2(3^{1/2})/9\pi$	$\approx$	12.25%
Cube	$2(31/2)/3\pi$	$\approx$	36.75%
Octahedron	$1/\pi$	$\approx$	31.83%
Dodecahedron	$3^{1/2}(5 + 5^{1/2})/6\pi$	$\approx$	66.49%
Icosahedron	$[2(5 + 5^{1/2})]^{1/2}/2\pi$	$\approx$	60.55%

We are next introduced to the Platonic chemistry, which is based on the *law of the conservation of triangles.*

> One part water (20 equilateral triangles) =
> one part fire (4 equilateral triangles) +
> two parts air (2 x 8 equilateral triangles).

> One part air (8 equilateral triangles) =
> two parts fire (2 x 4 equilateral triangles).

> Two parts water (2 x 20 equilateral triangles) =
> five parts air (5 x 8 equilateral triangles).

Now of course there are different types of earth, different types of fire, different types of water, and different types of air, but these can all be explained by the fact that in some cases the triangles are held together closely, in some cases loosely. In some cases the triangles are huge; in other cases tiny. All of physics and chemistry, everything that happens in the material universe, is to be explained in terms of the 30-60-90 and isosceles right triangles. This applies to the human body as well, and death is also to be explained in terms of a disease of the triangles of which the body is composed.

> Each individual comes into the world having a fixed span, and the triangles in us are originally framed with power to last for a certain time, beyond which no man can prolong his life. (89 b 7–c 4)

The dialogue is now over, and Plato leaves us with a happy ending.

> We may now say that our discourse about the nature of the universe has an end. The world has received animals, mortal and immortal, and is fulfilled with them, and has become a visible animal containing the visible—[the universe] is the sensible God who is the image of the intellectual, the greatest, best, fairest, most perfect—[It is] the one only begotten heaven. (92 c 4–9)

-tive See the entry **-ive**.

topological More correct would have been *topologic* since *-alis* is a Latin ending and the word is Greek. See **topology**.

topology The word was invented by Listing (1808–1882) to mean the *science* (-λογία) of *position* (τόπος). The Greeks had two "ologies," theology and astrology. The suffix is derived from the noun λόγος, which means *word, reason*; it does not refer to the noun λογία, which means the *collection of taxes or contributions*.

torsion The Latin verb *torquo, torquere, torsi, torsus* means *to twist, to torture by twisting the joints of the body*. The English word is derived from the late Latin noun *torsio, torsionis* formed from the fourth principal part of the verb. The torsion τ is the measure of the amount that a space curve twists at a point. It is defined by the equation

$$d\mathbf{B}/ds = -\tau \mathbf{N},$$

where $\mathbf{B}$ is the unit binormal, $\mathbf{N}$ is the unit normal, and s is arclength.

torus The Latin noun *torus* means *any round swelling or protuberance*.

total The Latin adjective *totus* means *whole*. The superfluous superimposition by the Scholastic philosophers of the adjectival suffix *-alis* produced the low word *totalis* with the same meaning.

totient The Latin adverb *totiens* means *so often, so many times*. The *s* was changed to *t* by people used to Latin third- and fourth-conjugation participles ending in *-iens, -ientis*.

trace This verb is derived from the fourth principal part of the Latin verb *traho, trahere, traxi, tractus, to drag*. It came into English from French, where the *t* had already been dropped. The *trace* of a matrix is the sum of the diagonal entries.

tractor The Latin verb *traho, trahere, traxi, tractus* means *to drag*. The frequentative verb *tracto, tractare* is formed from it with the meaning *to drag frequently*. The *tractor* is *the fellow who drags*.

tractory This is another name for the *tractrix*. It comes from the adjective in the expression *linea tractoria, the dragging curve*.

tractrix This noun is the feminine of *tractor* and means *a female who drags*. The feminine form is used because the Latin word for plane curve, *linea*, is feminine.

trajectory The Latin verb *traicio, traicere, traieci, traiectus* means *to throw* (*iacio*) *across* (*tra-* from *trans*). From the fourth principal part were formed the nouns *traiectio, traiectionis* and *traiectus, traiectūs* with the meaning *a passing or crossing over*. The addition of the adjectival suffix *-orius* to the stem of the latter noun produced the adjective *traiectorius, -a, -um* with the meaning *pertaining to a crossing*.

transcendental The prefix *trans-*, the Latin preposition *across*, should only be combined with words of Latin origin. With words of Greek origin, one should use the prefix *meta-* (μετά-). Thus, the theologians speak of the *metamorphosis* or *transfiguration*, but not of the *metafiguration* or the *transmorphosis*. For this reason, the word *transdermal* used by the drug companies is low. A transcendental number is a real number that is not the root of a polynomial with rational coefficients. Transcendental functions are functions that are not algebraic.

transfinite The use of the preposition *trans*, which means *across* in Latin, as a prefix to give the idea of *across, through,* or *over* to the following verb is common. There is, however, no verb *transfinio*. The adjective *transfinite* is a modern invention. The use of the prefix *trans-* with adjectives in English is common and acceptable when the intent is to take something to the next logical level. A transfinite cardinal number is a cardinal number that is not finite, such as $\aleph_0$, the cardinal number of the set of positive integers, and $\aleph$ or c, the cardinal number of the continuum.

transform The Latin verb *transformo, transformare, transformavi, transformatus* is used by Vergil to mean *to change the shape*; Ovid has the adjective *transformis* with the meaning *with changed shape*.

transformation The addition of the suffix *-io* to the stem of the fourth principal part of the verb *transformo* produced the noun *transformatio, transformationis* meaning *a change in form*.

transitive The Latin verb *transeo, transire, transivi, transitus* means *to go* (*eo*) *across* (*trans*). From the fourth principal part were formed the nouns *transitio* and *transitus*, each meaning *a passing over*. The addition of the adjectival suffix *-ivus* to the stem of each of them gives the adjective *transitivus*, meaning *passing over*, which was used by the grammarian Priscian of Caesarea in Palestine, who taught Latin at Constantinople (A.D. 500).

translate This word was formed from the fourth principal part of the Latin verb *transfero, transferre, transtuli, translatus*, which means *to carry across*. It began as a synonym for *transfer*, which was produced from the first principal part of the same verb.

translation The addition of the suffix *-io* to the stem of the fourth principal part of the verb *transfero* produced the noun *translatio, translationis*. A *translation of axes* of the Cartesian plane is a function that assigns to each point *(x,y)* a new pair of coordinates *(x', y')* such that $x = x' + h$ and $y = y' + k$ for some fixed pair of real numbers h and k.

transpose This verb was formed from the fourth principal part of the Latin verb *transpono, transponere, transposui, transpositus*, which means *to put (pono) over (trans), remove, transfer.*

transposition The Latin verb *transpono, transponere, transposui, transpositus* means *to put (pono) over (trans), remove, transfer.* The late addition of the suffix *-io* to the stem of the fourth principal part produced the technical noun *transpositio, transpositionis* with the meaning *an exchange.*

transverse The Latin adjective *transversus, -a, -um* means *oblique, athwart.* It is the fourth principal part of the verb *transverto, transvertere, transverti, transversus*, which means *to turn (verto) across (trans).*

transversal Though *transversus* is already an adjective, Albertus Magnus (1193–1280) superimposed the adjectival suffix *-alis* upon the stem to create the adjective *transversalis*, from which the English adjective was derived. Albert wrote the first original commentary on Euclid's *Elements of Geometry* in the Latin language.

trapezium This is the transliteration of the Greek noun τραπέζιον, the diminutive of τράπεζα, *table.*

trapezoid This means *table-shaped* in Greek. From the nouns τράπεζα, *table*, and εἶδος, *shape*, the Greeks formed the adjective τραπεζοειδής, τραπεζοειδές with the meaning *shaped like a table.* It modified the noun σχῆμα, *shape*, which was often understood, so that one just wrote or read τὸ τραπεζοειδές, *the table-shaped [figure].* As a result, though an adjective, it came to be treated in English as a noun.

trapezoidal This modern word is macaronic, formed by the addition of the Latin adjectival suffix to the stem of a Greek adjective. The method of *trapezoidal approximation* approximates the area of a plane region by inscribing it with trapezoids.

triad This is the stem of the Greek noun τριάς, τριάδος, which means *the number three*.

triangle This is the Greek word τρίγωνον, *triangle*. It is *a figure with three angles*, from the prefix τρι- (from τρεῖς, the number *three*) and γωνία, *angle, corner*. It was taken over into Latin as the noun, *triangulum*; the Latin language also has the adjective *triangulus, -a, -um*, with the meaning *three-cornererd*. The inequality $|a + b| \leq |a| + |b|$ is called the *triangle inequality*.

triangular The Romans made the Latin adjective *triangularis* from the noun *triangulum*.

triangulate Albertus Magnus (1193–1280) used the adjective *triangulatus*, which has the appearance of the past participle of a verb *triangulo, triangulare, triangulavi, triangulatus*, which has not been found in any work of literature. The English verb was created by adding the suffix *-ate* to the stem of the noun *triangulum*. The word has entered the vocabulary of the talking heads, for on the April 1, 2012, episode of Fareed Zaccaria's *Global Public Square*, guest Matt Franck said, "Let me try to triangulate between Jon [Meacham] and Sally [Quinn]." He was simply presenting a third point of view midway between theirs.

triangulation The addition of the Latin suffix *-atio* to the stem of the noun *triangulum* produced this word, the English form of the Latin *triangulatio, triangulationis* used by Abelard in the twelfth century.

trichotomy This means *a cutting up into three parts*. It is derived from the Greek τρίχα, *threefold*, and τομή, *a cutting*, from the verb τέμνω, *to cut*. The *law of trichotomy* for the set $\mathcal{N}$ of natural numbers is the property that for all $a, b \in \mathcal{N}$, $a < b$, $a = b$, or $b < a$.

trident This is the three-pronged staff of Neptune, from the Latin *tres, three*, and *dens dentis, tooth*.

trifolium A *trifolium* is something with three (*tres*) leaves (*folium* in the singular, *folia* in the plural). The plural is *trifolia*, though *trifoliums* can be tolerated (barely).

trigonometrical This English word was formed after the analogy of *geometrical*. It is centuries old, and has fortunately not displaced the better word *trigonometric*. The addition of the stem of the Latin adjectival suffix *-al* is pleonastic since *trigonometric* is already an adjective.

trigonometry The Greek noun τριγωνομετρία was transliterated into Latin as *trigonometria*. From thence it entered French as *trigonométrie*, and the ending *-ie* became *-y* in English as in the case *Marie, Mary*.

trihedral This adjective is derived from τρεῖς, the Greek word for *three*, and ἕδρον, the Greek word for *seat*. (The Greek diphthong ει was regularly transliterated by *i*, as were the letters ι and η.) There was then appended the Latin suffix *-alis* to produce an adjective. *Trihedric* would have been the correct form, but the offspring of ignorance has prevailed by immemorial custom. A *trihedral angle* is the solid angle formed by three planes coming together at a point, as at the vertices of a tetrahedron.

trilateral This is derived from *tres*, the Latin word for *three*, and *latus, lateris*, the Latin word for *side*. The suffix *-alis* was then added to the stem to produce an adjective *trilateralis* meaning *having three sides*.

trillion This is an absurd word for the number 1,000,000,000,000 or $10^{3 + 3(3)}$. Similarly, a billion is $10^{3 + 2(3)}$, a quadrillion is $10^{3 + 4(3)}$, a quintillion is $10^{3 + 5(3)}$, a sextillion is $10^{3 + 6(3)}$, etc.

trinomial See the entry *binomial* above. The Greeks added the prefix τρι- to a word to indicate three or three times. Thus, τρίγωνον is a figure with three corners.

trisect This verb was well-formed from the Latin prefix *tri-* and the verb *seco, secare, secui, sectus,* which means *to cut,* on the analogy of the previously existing word *bisect.*

trisector This is a Latin noun meaning *the one who cuts into three parts.* The prefix *tri-* is from *tres, three,* and *sector,* a noun of agent from *seco, secare, to cut.* A *trisector* is a fellow who persists in attempting to trisect an arbitrary angle by means of unmarked straightedge and compass alone despite the fact that modern science has shown that such a construction is impossible. He is not sufficiently educated to understand the proof that his labors are in vain. *Trisector* is therefore a pejorative term applied to people who are *busy* about an enterprise in the eighteenth-century sense of that word, uselessly active.

trisectrix This is the Latin feminine of *trisector.* The reason for the feminine gender is the same as in the cases of all names of curves; the name refers to *linea* understood, and *linea* is feminine. The *trisectrix of Maclaurin* is a curve with loop and asymptote. It is the pedal curve with respect to the origin of the parabola with equation $y^2 = -4a(x + 3a)$, where a is a positive parameter. When in standard form, its Cartesian equation is $y^2 = x^2(3a + x)/(a - x)$, and its polar equation is $r = a \sec \theta - 4a \cos \theta$. The line $x = a$ is the asymptote. The area of the region inside the loop and the area of the region intercepted between the trisectrix and its asymptote are the same, $3^{3/2}a^2$.

The trisectrix may be produced in the following manner. Let ℓ_1 and ℓ_2 be two straight lines rotating counterclockwise about the points *(0,0)* and *(2a,0)* with uniform angular speeds ω_1 and $3\omega_1$, respectively. Assume that at time zero, both are horizontal. Then Maclaurin's trisectrix is the locus of intersection of the two lines. The trisectrix produced in this manner is the reflection across the *y*-axis of the one whose equation is given in the paragraph above.

trivial *Tres* is the Latin word for *three,* and *via* means *road. Trivia* is therefore a branching into *three roads.* The addition of the adjectival ending *-alis* produces *trivialis,* meaning *that which pertains to the intersection of three roads.* The *trivium* was the set of three non-mathematical subjects that together with the mathematical *quadrivium* made up the seven liberal arts. The trivium consisted of grammar, logic, and rhetoric. Grammar meant Latin, the major language of Western civilization and the only language in Western Europe for 1,500 years that had a literature. Logic, as John Stuart Mill said, "clears up the fogs which make us believe that we understand a subject when we do not." (This statement is found in his inaugural address as rector of the University of St. Andrews, 1867.) Rhetoric means how to stand up before an audience and speak well, but to do this is impossible without first knowing something and having a command of the works of the best authors. The fact that the adjective *trivial* descended into a pejorative term is telling. Paul Halmos once asked a student in my presence what subjects he was studying. The student replied, "Mathematics, physics, and Greek." Halmos commented, "One of those is difficult."

trochoid This is the name of the plane curve that is the locus of a fixed point a distance b from the center of a circle of radius r rolling without slipping on a straight line. If $b < r$, the trochoid is called *curtate*; if $b > r$, the trochoid is called *prolate*. If $b = r$, the trochoid is a *cycloid*. This word is the juxtaposition of the two Greek words τρόχος, *wheel*, and εἶδος, *shape*. The curve of which it is the name does not, however, look like a wheel; rather, its construction is accomplished by the use of a wheel. The name is therefore a misnomer. The parametric equations of the trochoid are

$$x = r\theta - b \sin \theta$$
$$y = r - b \cos \theta.$$

The area of the plane region under one arch of the trochoid, $0 \le \theta \le 2\pi$, is $\pi r(3b^2 + 2r^2)$. See the entries **curtate**, **prolate**, and **cycloid**.

truncate The Latin adjective *truncus, -a, -um* means *lopped, maimed, mutilated, cut short*. From it was formed the verb *trunco, truncare, trucavi, truncatus* with the meaning *to shorten by cutting off*. The English verb was taken from the fourth principal part of the Latin verb.

tunnel Weekley says that there is a medieval Latin word *tunna*, which in French was *tonne* with diminutive *tonnel*; *un tonnel* was a cylindrical cask. Tunnel problems were first considered by Adelard of Bath (twelfth century), the first Latin translator of the Arabic editions of Euclid. Suppose a linear tunnel is dug from a point on the surface of the earth through the center to the antipode. If an object of mass m is dripped into this tunnel, what will happen to it? Adelard argued in his *Quaestiones Naturales* that the object must fall to the center of the earth and remain there, but under Newton's law of gravity, the point mass is required to oscillate in simple harmonic motion between the antipodes. The period of the oscillation, that is, the time required for the mass to return to the point whence it was dropped, is one hour, twenty-four minutes, twenty-nine seconds. Newton showed that if one wanted to travel under the force of gravity in the shortest time possible from a point A on the earth's surface to some other point B on the earth's surface whose distance from A along the great circle connecting them is s, then the best of all possible tunnels from A to B is that of the hypocycloid with neighboring cusps at A and B, produced by rolling inside the great circle a smaller one of circumference s.

type I , type II error The Greek noun τύπος means *a blow* and then *the mark left by a blow*. The Latin noun *error* means *a mistake* and is derived from the verb *erro, errare, to wander*. The use of numbers in definitions, such as first category, second category, type I, type II, etc. is acceptable when no more suggestive terminology is conceivable.

U

ultrafilter The Latin preposition *ultra* means *beyond*. The noun *filter* is from the medieval Latin *filtrum*, which means *felt*. See the entry **filter**. An *ultrafilter* $\mathbb{B}$ for a set X is a filter for X such that for each $A \subseteq X$, either $A \in \mathbb{B}$ or $X - A \in \mathbb{B}$ but not both.

umbilical *Umbilīcus* is the Latin word for *the navel*; it is related to the Greek noun of the same meaning ὀμφαλός. The Latin adjective *umbilicaris*, with the meaning *pertaining to the navel*, was formed by adding the adjectival suffix *-aris* to the stem of the noun; in medieval times there arose the additional adjective *umbilicalis* with the same meaning. An *umbilical point* of a surface $\mathcal{S}$ is a point P on $\mathcal{S}$ that is either a circular point or a planar point of $\mathcal{S}$; in the former case, the surface near P resembles a sphere, and in the latter case it resembles a plane there. See James and James.

umbra This is the Latin word for *shade* or *shadow*.

unconditional The prefix *un-* is the Germanic equivalent of the Greek *alpha privativum* ἀ- and the Latin *in-*; it negates the adjective to which it is adjoined. It should not be added to words of Latin origin, for which the prefix *in-* is appropriate. An *unconditional or absolute inequality* is an inequality that is true for all values of the variable, for example, $x^4 + 1 > x$. See the entry **conditional**.

undecidable See the entry **decidable**. *Undecidable propositions* are propositions that cannot be proven true or false.

undecillion This is an absurd word supposed to mean $10^{6 + 11(3)}$. It is less comical to say "ten to the thirty-ninth." *Undecem* is the Latin word for *eleven*; *-illion* is derived from the Latin *mille*, *a thousand*, with the augment *-on* (originally *-one* in Italian), which means *big*. A million is just a big thousand.

undefined The Germanic negative prefix *un-* has been added to the word *defined* of Latin origin to produce the hybrid *undefined*. It would have been better to say *indefined* as we say *indefinite*, but it is too late now. *Defined* is from the verb *definio, definire, definivi, definitus*, which means *to set the boundaries*. The plural noun *fines* in Latin means *enclosed area, territory*. The force of the prefix *de-* is to add the sense of thoroughness to the action.

undetermined The Germanic negative prefix *un-* has been added to the word *determined* of Latin origin to produce the hybrid *undetermined*. *Indetermined* would have been the correct formation, in the same way as we say *indeterminate*. *Determined* is derived from the Latin verb *determino, determinare, determinavi, determinatus*, which means *to set the boundaries*. The noun *terminus* in Latin means *the boundary mark*. In the theory of differential equations, the method of undetermined coefficients is a method of determining a particular solution of a linear non-homogeneous differential equation. One guesses that the solution must be of a certain form with certain parameters and then proceeds to determine the parameters. The force of the prefix *de-* is to give the sense of thoroughness to the action.

uni- This Latin prefix is derived from the adjective *unus*, which means *one*. It is added to other adjectives of Latin origin to give the meaning of *sole, unique*. It cannot be added to adjectives derived from other languages without producing a low word. The Greek equivalent is μονο-, *mono-*.

unicursal This nineteenth-century adjective is the combination of the Latin particle *uni-* and the medieval Latin adjective *cursalis*. (The proper form of the adjective is the ancient *cursualis*.) The Latin verb *curro, currere, cucurri, cursus* means *to run*. From its fourth principal part is derived the noun *cursus, cursūs, running*. The adjective *unicursal* is applied to paths in space and is a synonym for *continuous*; the idea is that the path can be drawn *with one course* of the pen.

uniform The Latin adjective *uniformis* means *having one form, simple*.

uniformity The Latin noun *uniformitas, simplicity*, came into French as *uniformité* , which then became the English *uniformity*.

unilateral The Romans added the prefix *uni-* to an adjective to express the restriction *only* or *one*. The formation *unilateralis* is a modern word meaning *pertaining to one side* (*latus*), but it is well made on the ancient model.

unimodal This is a modern word used in the theory of probability. A *unimodal random variable* is one that has one and only one mode. See the entry **mode**.

union From the Latin adjective *unus, -a, -um, one*, came the verb *unio, unire, univi, unitus, to bring together into one*, and the noun *unitas, unitatis, oneness*. From the Latin noun came the French noun *unité*, which was transformed into the English *unity*.

unipotent A square matrix M is *unimodal of order n* if n is the least positive integer such that $M^n = I$. The word is low and not at all suggestive of its technical meaning.

unique This is the French metamorphosis of the Latin adjective *unicus, only*. It came into England with the Normans.

unit From the Latin adjective *unus, -a, -um, one*, came the verb *unio, unire, univi, unitus, to bring together into one*, and the noun *unitas, unitatis, oneness*. From the Latin noun came the French noun *unité*, which was transformed into the English *unity*.

unital As far as we know, there never was any Latin adjective *unitalis*. The English word has existed for centuries with the meaning *unitary*. A *unital* module is a module whose ring of scalars contains a multiplicative identity.

unitary See the entry **unit**. The Latin adjective *unitarius* is modern and was coined on the analogy of *Trinitarian* to describe those who denied three persons in one God. Suppose an inner product *(x,y)* is defined on a vector space Υ. Then a transformation U is *unitary* on Υ if $(x,y) = (Ux,Uy)$ for all $x, y \in \Upsilon$. The name comes from the fact that a transformation on a vector space is unitary if and only if it preserves the length of each unit vector.

unity See the entry **unit**. De Moivre's formula for the n roots of unity is $cos(2\pi j/n) + i \, sin(2\pi j/n), j = 1, 2, 3,\ldots,n$.

universal The Latin adjective *universalis* was used by Quintillian, Pliny, and Livy with the meaning *general, pertaining to the whole*. The Latin adjective was formed by adding the adjectival suffix *-alis* to the stem of the noun *universum* (see the following entry).

universe This comes to us through French from the Latin adjective *universus*, which means *combined in one whole, entire;* the neuter singular *universum* was used as a noun meaning *the whole world*. It was formed by the combination of the prefix *uni-* with the past participle *versus* of the verb *verto, vertere*, which means *to turn*.

unknown This noun and adjective is compounded of the Germanic negating prefix *un-* and the Latin verb *[g]nosco, [g]noscere, [g]novi, [g]notum*, which means *to know*; the initial *g* had dropped out of this verb by the classical age. The word is cognate with the Greek verb γιγνώσκω of the same meaning.

usual The Latin verb *utor, uti, usus* means *to use*. From its third principal part comes the adjective *usualis, -e* whence our adjective *usual*. The *usual topology* on a metric space is the topology in which a set is open if and only if it is the union of open spheres.

V

valence The Latin verb *valeo, valere, valui, to be strong,* has the present participle *valens, valentis* with the meaning *powerful.* From the participle came the late Latin noun *valentia, power,* from which the French (and English) noun is derived. Finkbeiner and Lindstrom (p. 225) define the *valence* or *degree* of a vertex *v* in a graph *G* to be the number of edges of *G* that are incident at *v*.

valid From the Latin verb *valeo, valere, valui,* which means *to be strong,* is derived the adjective *validus* with the meaning *strong, powerful.* The English adjective is produced by dropping the nominative case ending *-us* from the Latin adjective.

valuation The late Latin noun *valuatio, valuationis* means *worth* and is derived from the verb *valeo, valere, valui,* which developed the meaning *to be worth* from its original meaning *to be strong.* If $(X, \mathcal{J})$ is a topological space, a *valuation* is a mapping f from $\mathcal{J}$ to $[0, \infty]$ satisfying i) $f(\emptyset) = 0$, ii) if A and B are open sets such that $A \subseteq B$, then $f(A) \leq f(B)$, and iii) if A and B are any open sets, then $f(A \cup B) + f(A \cap B) = f(A) + f(B)$.

value The French verb *valoir* means *to be worth*; its past participle is *valu,* from which is derived the French and the English noun *value.* The French verb is derived from the Latin verb *valeo, valere, valui,* which means *to be strong.*

vanish The Latin adjective *vanus* means *empty.* According to Lewis and Short, its etymology is uncertain, though they call the attention of their readers to the verb *vaco, vacare,* which means *to be empty.* The ending *-us* became *-ish* through the mispronunciation of the unlearned. The modern technical use of the word as a verb meaning *to be zero* (of functions) dates back to at least the beginning of the eighteenth century.

variable *Vario* is a first-declension Latin verb that is transitive in the sense of *to diversify, to alter,* and intransitive in the sense of *to be different, to vary.* From this there developed the medieval Latin adjective *variabilis* with the meaning *to be changeable,* whence we get our technical term *variable.* The use of *variable* as a noun in the modern mathematical sense is at least as old as the early nineteenth century.

variance From the Latin adjective *varius,* which means *manifold,* came the verb *vario, variare, variavi, variatus* with the meaning *to diversify;* the verb's present participle is *varians, variantis.* From this participle was formed the noun *variantia, difference,* used by Lucretius; it is the origin of the French, and of the English, *variance.*

variate This English noun is derived from the fourth principal part of the Latin verb *vario.* See the previous two entries. A *random variate* is an element of the range of a random variable.

variation From the Latin adjective *varius,* which means *manifold,* came the verb *vario, variare, variavi, variatus* with the meaning *to diversify.* From the fourth principal part of the verb came the noun *variatio, variationis,* used by Livy. This noun is the origin of the English *variation.* The calculus of variations is that branch of the theory of differential equations that finds the curves that solve certain maxima and minima problems. The method of variation of parameters is due to Lagrange (1736–1813); it is employed to find a particular solution of a non-homogeneous differential equation $y'' + by' + cy = 0$ when the method of undetermined coefficients fails.

vault This word is the corruption of the fourth principal part of the Latin verb *volvo, volvere, volvi, volutus,* which means *to roll.* A *barrel vault* is a ceiling in the shape of a half cylinder, the half on one side of a dividing plane through the axis. The nave of St. Peter's Basilica has the most famous of barrel vaults. The *Roman vault* is the top half of the solid of intersection of two equal circular cylinders intersected at right angles; I have seen the whole solid of intersection called a *bicylinder,* and the word is acceptable. The Romans used this type of vault in their public baths. Brunelleschi used the top half of the solid of intersection of four equal right circular cylinders intersected at 45° angles for the dome of the cathedral of Florence. Michelangelo used the top half of the solid of intersection of eight equal right circular cylinders intersected at 22 ½° for the dome of St. Peter's, the most famous work of art in the world. In general, we take a cylinder of radius r and consider the vault that is the top half of the solid of intersection of all the cylinders produced by rotating the given one about a fixed line perpendicular to its axis by $i(2\pi/n)$, $i = 1, 2, 3,..., n/2$, n an even positive integer. If $n = 4$, this is the Roman vault; if $n = 8$, it is the Brunelleschi vault; and if $n = 16$, it is the Michelangelo dome. In all cases, the curves of intersection (the "ribs") are elliptical arcs. The volume of the vault is $[2n \ tan(\pi/n)]r^3/3$, and the centroid is $3r/8$ up the axis, the same as in the case of the hemisphere.

vector The Latin verb *veho, vehere, vexi, vectus* means *to carry*. By adding the suffix *-or* to the stem of the fourth principal part one produces the agent, *the one who carries*. The use of the word as a mathematical technical term is as least as old as Hamilton.

vectorial By adding the adjectival suffix *-alis* to the noun *vector* one produces the adjective *vectorialis* with the meaning *having to do with a carrier*. The English adjective is produced by dropping the nominative case ending *-is*.

velocity The Latin adjective *velox, velocis* means *fast*. By adding the nominal suffix *-itas* to the stem one produces the noun *velocitas* with the meaning *speed*.

versed sine The adjective *versed* is derived from the fourth principal part *versus* (meaning *turned*) of the Latin verb *verto, to turn*. If a central angle θ is inscribed in a circle of radius r, the *sinus versus* or *versed sine of* θ is what we would call $r(1 - cos\ \theta)$. The definition given in the *Oxford English Dictionary*, that the versed sine was "originally the segment of the diameter intercepted between the foot of the sine and the extremity of the arc," still holds today if $r = 1$. The versed sine was also called the *sagitta* or arrow, the arc being the bow and twice the sine being the string.

versiera This is the name of a curve discussed by Maria Gaetana Agnesi in her book *Instituzioni analitiche* of 1748; by *versiera* Agnesi intended a reference to the versed sine, for the curve has the parametric equations $x = a \cot \theta$, $y = a(1 - \cos 2\theta)/2$, but the word is also the abbreviated Italian name for a witch (*avversiera* = adversary). Its Cartesian equation is $y = a^3/(x^2 + a^2)$, where a is a positive constant. For one value of its parameter a, it appears in probability as the probability density function of the random variable with Student's *t*-distribution with one degree of freedom. B. Williamson, in his book *Integral Calculus* published in 1875, section vii, page 173, has perhaps the first mention of the curve as *the witch* in English:

> Find the area between the witch of Agnesi $xy^2 = 4a^2(2a - x)$ and its asymptote. (Taken from the entry in the *Oxford English Dictionary*)

vertex This is a Latin noun that means *something that turns*, from *verto, vertere, verti, versus, to turn*, for example, *a whirl, an eddy of water, a whirlwind or gust of wind*. It then came to mean the *crown of the head, the summit of anything*. Its use to mean the special points on curves such as the conic sections or surfaces such as the Eulerian quadric surfaces is as least as old as the sixteenth century.

vertical The adjectival suffix *-alis* was added to the stem of the noun *vertex, verticis* to form the adjective *verticalis*, from which the English adjective was made by removing the nominative case ending *-is*. See the entry **vertex**.

vibration The Latin noun *vibratio, vibrationis*, was formed from the fourth principal part of the verb *vibro, vibrare, vibravi, vibratus*, which means *to brandish, to move rapidly to and fro*.

vigintillion This is an absurd word concocted by throwing together the Latin word for *twenty*, *viginti*, and the ending *-illion*, stupidly fashioned from the Latin word *mille*, which means *a thousand*. It is supposed to be the name of the number $10^{3\,+\,20(3)}$. It is not necessary that everything have a name of one word. A better name for the number in question is, "ten to the sixty-third" or, for the unmathematical student, "one with sixty-three zeroes after it."

vinculum From the Latin verb *vincio, vincire, vinxi, vinctus, to bind or tie around*, was formed the noun *vinculum*, which means *a band, cord, or chain*. The *vinculum* is a bar written above an algebraic expression, equivalent to enclosing that expression within parentheses. It is old-fashioned and not recommendable anymore.

virgule The Latin noun *virga* means *a green twig*. By addition of the suffix *-ula* to the stem one produces the diminutive *virgula, a little twig*. The *virgule* is the name of the symbol / in expressions such as *and/or* and *his/her*. It is poor style and should never be used.

virtual This adjective has become the most hideous computer lingo, as in *virtual channel, virtual function, virtual memory, virtual reality*. The Latin noun *vir* means *man*, and the related noun *virtus, virtutis* means *manliness, manly excellence, strength*. The addition of the adjectival suffix *-alis* to the stem results in the late Latin adjective *virtualis*, with the meaning *pertaining to manly excellence*.

volume This is the Latin noun *volumen, something rolled up*. It is derived from the verb *volvo, volvere, volvi, volutus, to roll*. It came to mean a book since most ancient books were scrolls of parchment or papyrus that were sewed together one after another and then rolled up. (*Codex*, which means *tree trunk*, was a book made of wooden tablets, one put on top of another and then later of pages of velum or papyrus laid one on top of another and sewed together in that way.) The use of *volume* to mean *size* or *bulk* is at least as old as the sixteenth century.

vulgar From the Latin noun *vulgus, the multitude*, was derived the adjective *vulgaris*, with the meaning *common, ordinary, usual*. The English word was produced by removing the nominative masculine case ending *-is*.

W

word This noun is cognate with the Latin *verbum*, which has the same meaning. *Verbum*, in turn, is related to the Greek εἴρω, which means *to say, speak, or tell.*

X

x-axis The use of the letters x, y, and z for unknown quantities is due to Descartes, who used the letters a, b, and c for known quantities. The *Oxford English Dictionary* says that there is no evidence to support the hypothesis that x is actually the metamorphosis of the medieval cursive r, the first letter and abbreviation of the noun *res*, *thing*. (Medieval authors writing in Latin referred to an unknown quality as *res*, that is, *the thing*. To abbreviate this reference, they used r, the first letter of that word.)

Y

y-axis The letter y is neither Latin nor Greek. Dr. Johnson says in his dictionary that it was a symbol much used by the Saxons instead of i. When transliterating Greek into English, y is to be used for *upsilon*.

Z

z-axis The letter *z* is the Greek *zeta*. It is not a Latin letter and was used by the Romans only to transliterate the Greek *zeta* in words that they took over from the Greek language.

zenith This is the metamorphosis of the Arabic construct phrase سمت الرأس, which means *the way overhead*, from الرأس, *the head*, and سمت, *way*.

zero The infinitive *ṣifr* of the Arabic verb صفر, which means *to be empty*, became the *zephyrum* of the medieval Latin translators, and *zephyrum* was mutilated into *zero*.

Bibliography

The presentation of the bibliography provides the opportunity for me to condemn the practice of referring to items in bibliographies by the notation [B-G 10], [Cap 03], [Chev 58], [Fabre 08], etc., that is to say, by the often comical abbreviation of the author's name followed by the last two digits of the year of publication. It is deplorable that such an illiterate system is becoming common enough to require reprobation. When it is necessary to make frequent references to items in a bibliography, no one has yet improved upon the tried and true method of numbering the entries consecutively in accordance with alphabetical order and then referring to them by author or entry number followed by page number. Also eccentric is the substitution of the title *References* for *Bibliography*.

Alford, Henry, *A Plea for the Queen's English: Stray Notes on Speaking and Spelling*, tenth thousand, Alexander Strahan, publisher, London and New York, 1866.

Allen and Greenough's New Latin Grammar for Schools and Colleges founded on Comparative Grammar, edited by J. B. Greenough, A. A. Howard, G. L. Kittredge, Benj. L. D'Ooge, Ginn and Company, 1916.

Brown, Francis, Driver, S. R., and Briggs, Charles A., *A Hebrew and English Lexicon of the Old Testament, with an Appendix Containing the Biblical Aramaic, Based on the Lexicon of Wilhelm Gesenius as Translated by Edward Robinson*, Edited with constant reference to the Thesaurus of Gesenius as completed by E. Rödiger, and with authorized use of the latest German editions of Gesenius's *Handwörterbuch über das Alte*

Testament by Francis Brown with the co-operation of S. R. Driver and Charles A. Briggs, Oxford at the Clarendon Press, 1968.

Cajori, Florian, *A History of Mathematical Notations*, two volumes in one, Dover Publications, Inc., New York, 1993.

Calinger, Ronald (editor), *Classics of Mathematics*, Prentice-Hall, Englewood Cliffs, New Jersey, 1995.

Cassell's New Latin Dictionary, revised by D. P. Simpson, M.A., Funk & Wagnalls Company, New York, 1959.

Crystal, David, *The Cambridge Encyclopedia of the English Language*, Cambridge University Press, 1995.

Egger, Karl, *Lexicon Nominum Virorum et Mulierum*, second edition, Editrice Studium, Rome, 1963.

Fowler, H. W., *A Dictionary of Modern English Usage*, second edition revised by Sir Ernest Gowers, Oxford University Press, New York and Oxford, 1965.

Freytag, George William, *Georgii Wilhelmi Freytagii Lexicon Arabico-Latinum praesertim ex Djeuharii Firuzabadiique et aliorum arabum operibus, adhibitis Golii quoque et aliorum libris confectum, accedit index vocum latinarum locupletissimus*, four volumes, Halis Saxonum (Halle), apud C. A. Schwetschke et filium, 1830.

Goodwin, William W., *A Greek Grammar*, Macmillan, St. Martin's Press, New York, 1968.

Hale, William Gardner, and Buck, Carl Darling, *A Latin Grammar*, Ginn & Company, Publishers, The Athenaeum Press, Boston and London, 1903.

Haywood, J. A., and Nahmad, H. M., *A New Arabic Grammar of the Written Language*, revised edition, Harvard University Press, Cambridge, Massachusetts, 1965.

James, Glenn, and James, Robert C. (editors), *Mathematics Dictionary*, Students edition, D. Van Nostrand Company, Inc., Princeton, New Jersey, 1964.

Johnson, Samuel, *A Dictionary of the English Language: In Which The Words are deduced from their Originals, And Illustrated in their Different Significations By Examples from the Best Writers. To Which Are Prefixed, A History of the Language, And An English Grammar*, Printed by W. Strahan, for J. & P. Knapton et al., two volumes, London, 1755.

Knopp, Konrad, *Theory of Functions*, two parts, translated by Frederick Bagemiehl, Dover Publications, Inc., New York, 1945.

Lane, Edward William, *An Arabic-English Lexicon*, in eight parts, Librairie du Liban, Beirut, Lebanon, 1980. This is a photographic reproduction of the 1863 edition published in London by Williams and Norgate.

Lawrence, J. Dennis, *A Catalogue of Special Plane Curves*, Dover Publications, Inc., New York, 1972.

Lewis, Charlton T., and Short, Charles, *A Latin Dictionary founded on Andrews' edition of Freund's Latin Dictionary*, revised, enlarged, and in great part rewritten by Charlton T. Lewis and Charles Short, Oxford at the Clarendon Press, 1969.

Lewis, L. W. P., and Styler, L. M., *Foundations for Greek Prose Composition*, Heinemann Educational Books Ltd., London, 1968.

Liddell, Henry George, and Scott, Robert, *A Greek-English Lexicon, Revised and Augmented throughout by Sir Henry Stuart Jones with the Assistance of Roderick McKenzie, with a Supplement edited by E. A. Barber with the Assistance of P. Mass, M. Scheller, and M. L. West*, Oxford at the Clarendon Press, 1968.

Liddell, Henry George, and Scott, Robert, *A Lexicon Abridged from Liddell and Scott's Greek-English Lexicon*, Oxford at the Clarendon Press, 1963.

Lockwood, E. H., *A Book of Curves*, Cambridge University Press, Cambridge, 1961.

North, M. A., and Hillard, A. E., *Greek Prose Composition for Schools*, Rivingtons, London, 1965.

Oxford English Dictionary, *The Compact Edition of the Oxford English Dictionary*, complete text reproduced micrographically, twenty-seventh printing in the U. S., 2 volumes, Oxford University Press, April 1988.

Royden, H. L., *Real Analysis*, second edition, Macmillan Company, Collier-Macmillan Limited, London, 1970.

Schwartzman, Steven, *The Words of Mathematics: An Etymological Dictionary of Mathematical Terms Used in English*, Mathematical Association of America, 1994.

Smith, David Eugene, *History of Mathematics*, two volumes, Dover Publications, Inc., New York, 1958.

Smith, David Eugene, *A Source Book in Mathematics*, Dover Publications, Inc., New York, 1959.

Smyth, Herbert Weir, *A Greek Grammar for Colleges*, American Book Company, New York, 1920.

Struik, D. J. (editor), *A Source Book in Mathematics 1200–1800*, Princeton University Press, Princeton, New Jersey, 1986.

Thatcher, G. W., *Arabic Grammar of the Written Language*, fifth printing, Frederick Ungar Publishing Co., New York, 1976.

Thurston, Herbert, S.J., "American Spelling," *The Nineteenth Century*, vol. 60, no. 356 (October 1906), pages 606–617.

Webster's New Twentieth Century Dictionary of the English Language, unabridged second edition, The World Publishing Company, Cleveland and New York, 1962.

Weekley, Ernest, *An Etymological Dictionary of Modern English*, two volumes, Dover Publications, Inc., New York, 1967. This is an unabridged and unaltered republication of the work originally published by John Murray in London in 1921.

Wehr, Hans, *A Dictionary of Modern Written Arabic (Arabic-English)*, edited by J. Milton Cowan, fourth edition considerably enlarged and amended by the author, Otto Harrassowitz, Wiesbaden, 1979.

Wright, W., *A Grammar of the Arabic Language*, translated from the German of Caspari and edited with numerous additions and corrections by W. Wright, LL.D., third edition revised by W. Robertson Smith and M. J. de Goeje, two volumes, Cambridge at the University Press, 1967.

Yates, Robert C., *A Handbook of Curves and Their Properties*, J. W. Edwards, Ann Arbor, 1952.